Practical Design Patterns for Java Developers

Hone your software design skills by implementing popular design patterns in Java

Miroslav Wengner

BIRMINGHAM—MUMBAI

Practical Design Patterns for Java Developers

Group Product Manager: Gebin George
Publishing Product Manager: Kunal Sawant
Senior Editor: Rohit Singh
Technical Editor: Maran Fernandes
Copy Editor: Safis Editing
Project Coordinator: Prajakta Naik
Proofreader: Safis Editing
Indexer: Subalakshmi Govindhan
Production Designer: Prashant Ghare
Marketing Coordinator: Sonia Chauhan

First published: February 2023
Production reference: 1130123

Published by Packt Publishing Ltd.
Livery Place
35 Livery Street
Birmingham
B3 2PB, UK.

ISBN 978-1-80461-467-9

www.packt.com

I remember the days when I started exploring the Java platform at Sun Microsystems. It became not only a daily job but also a lifelong hobby and passion. This book is dedicated to my incredible wife, Tanja, my wonderful children, Maxi and Elli, and my entire family for giving me the energy, inspiration, and motivation to keep up the strength needed to complete this book.

– Miroslav Wengner

Foreword

It was November 2021. In a conversation about his developer career, a senior Java developer told me:

"I've been a senior engineer for almost 20 years. I don't know what to do to get beyond that level."

That is a typical statement. The length of time varies, but far too many developers and engineers feel stuck in a senior position.

Don't get me wrong, it's not bad to be a senior!

You may be working on great projects and using cool technologies. You are probably involved in solving sophisticated problems and facing deep technical challenges.

But there may come a day that you, as the aforementioned developer, feel like you could be doing more in your career, could have more influence in the direction of projects, and have independence and autonomy in your work, or maybe you would feel more fulfilled by inspiring and mentoring other developers.

Although this is something I have heard many times before, that particular instance got me thinking. You see, earlier that same day, I was looking at the results of the **Java Community Process (JCP)** election. Miroslav Wengner, my friend Miro, had just been elected to the Executive Committee of the premiere standard body of Java technology.

It was impossible not to think about the parallels...

I first met Miro when we worked together on the NetBeans team at Sun Microsystems, the company that created Java. Like the aforementioned developer, Miro had also worked with Java for many years. By the time I had a long conversation with him about his developer career, Miro too had been a senior engineer for a number of years. But he didn't stop there. Now, just a few years later, Miro has gone way beyond senior: freelancer, speaker, open source developer, committer of the Java Mission Control project at OpenJDK, Java Champion, and JCP EC Member, and why not a book author!

Like Miro, you don't have to be stuck at the senior level. You too can build your technical reputation, grow beyond senior positions, and have a larger impact on your project, company, and even the world.

All that leads us to this book that you hold in your hands.

What attracted me to Miro's book is the importance of design patterns in achieving exactly this goal: growing beyond the senior level. Whether you want to become a manager, continue your technical career as a staff-plus engineer, or even if you prefer to be more independent and grow as a freelancer or entrepreneur, you will need to take on more responsibilities and go beyond the code.

Although design patterns are directly connected to code, they are also separate and independent. They encapsulate proven solutions to the common challenges of designing quality software. Design patterns hover above a specific piece of code or even a specific project, and create a common vocabulary for explaining problems and solutions. Anyone in the business of building software should master this vocabulary to understand, communicate, and participate in discussions. After all, growing beyond senior is taking responsibility for your career. You start as a senior in your project, and then you expand your influence by getting involved with more impactful things.

When you master the vocabulary of design patterns, you are more equipped to go beyond your immediate project and get involved in other projects across your company and the industry. You'll maybe even get to the point of helping define the directions of technology through open source projects and foundations or through standards organizations, like Miro is doing inside the JCP.

This book will help you do exactly that. With a broad view of design patterns applied to the Java ecosystem, you will see not only the concepts and the vocabulary, but also the real implementation and impact of design patterns in software that you use daily.

And one of the things that always impresses me is Miro's ability to connect software with real things. His open source project, **Robo4J**, which even won a Duke's Choice Award, helps you turn your Java code into robots and drones capable of navigating the real world. I was delighted to see Miro taking the same approach to design patterns. The examples in this book, using multiple vehicles and their parts, are written using the latest Java 17+ features, bringing the designs to life and connecting them to real, concrete problems.

So, get started on your journey of building your reputation, growing your career, and reaching beyond the senior level. Knowing design patterns will help you fit right in, speak the language, and participate in the most important decisions. Let the amazing Miro be your guide, and feel free to reach out to me if you would like (as Miro once did) to talk about the next steps in your career.

Bruno Souza

Principal Consultant, Java Champion, JCP Executive Committee Member

@brjavaman

https://java.mn

Contributors

About the author

Miroslav Wengner is an engineer with a passion for resilient distributed systems and product quality. He is a co-author and contributor to the Robo4J project (a reactive soft real-time framework for robotics/IoT). Miro contributes to OpenJDK and participates in other open source technologies. He uses his passion for helping build resilient and scalable solutions.

Miro was selected for the Java Champions Program, recognized as a JavaOne Rockstar, and elected to the **Java Community Process (JCP)** as an executive committee member.

In addition to his day-to-day duties as a principal engineer at OpenValue, he shares his knowledge at conferences (JavaOne, Devoxx, and so on) and in blogs. Miro believes in the Java ecosystem and helps move it forward!

About the reviewer

Werner Keil works in areas such as Agile, BDD, Cloud-native DevOps, Java, Java EE/Jakarta EE, IoT, security, and microservices, helping Global 500 clients across various industries and IT vendors. Having worked for over 30 years as PM, coach, software architect, and consultant for different sectors, Werner is an Eclipse and Apache Committer and a JCP member in JSRs. Werner has won multiple JCP awards, including Member of the Year and Outstanding Spec Lead, and was recognized as Speaker of All Times by Java2Days, a large Java conference held in eastern Europe. He is an Eclipse Babel Language Champion, a project lead of Eclipse UOMo, and a committer member in the Jakarta EE Specification Committee.

Table of Contents

Part 2: Implementing Standard Design Patterns Using Java Programming

3

Working with Creational Design Patterns   65

4

Applying Structural Design Patterns   101

5

Behavioral Design Patterns   139

Part 3: Other Essential Patterns and Anti-Patterns

6

Concurrency Design Patterns   179

7

Understanding Common Anti-Patterns   213

Assessments   227

Index   231

Other Books You May Enjoy   240

Preface

The Java language is a tool for communicating with a very rich platform that provides many features ready to serve for application development. This book explores the latest developments in improving language syntax with examples of the most useful design patterns. The book reveals the relationship between features, patterns, and platform efficiency through example implementations. The book explores how theoretical foundations help improve the maintainability, efficiency, and testability of source code. The content helps the reader solve different tasks and provides guidance on how to approach programming challenges using a variety of sustainable and transparent approaches.

Who this book is for

This book is dedicated to all "hungry" engineers who want to improve their software design skills with new language enhancements and a closer look at the Java platform.

What this book covers

Chapter 1, *Getting into Software Design Patterns*, introduces us to the initial foundations of source code design structure and outlines the principles that should be followed to achieve maintainability and readability.

Chapter 2, *Discovering the Java Platform for Design Patterns*, discusses the Java platform, which is a very broad and powerful tool. This chapter exposes the features, functions, and design of the Java platform in more detail to continue building the foundation needed to understand the purpose and value of using a design patterns.

Chapter 3, *Working with Creational Design Patterns*, explores object instantiation, which is a key part of any application. This chapter describes how to approach this challenge while keeping the requirements in mind.

Chapter 4, *Applying Structural Design Patterns*, shows how to create source code that allows for clarity of relationships between required objects.

Chapter 5, *Behavioral Design Patterns*, explores how to create source code that allows objects to communicate and exchange information while maintaining a transparent form.

Chapter 6, *Concurrency Design Patterns*, discusses the Java platform and how it is a concurrent environment by nature. It shows how to harness its power for the designed application's purposes.

Chapter 7, Understanding Common Anti-Patterns, deals with anti-patterns that can be found in any application development cycle. It will help you deal with the root causes and their identification, and suggests possible anti-pattern remedies.

To get the most out of this book

To execute the instructions in this book, you'll need the following:

Software/hardware covered in the book	Operating system requirements
Java Development Kit 17+	Windows, macOS, or Linux
Recommended IDE VSCode 1.73.1+	Windows, macOS, or Linux
A text editor or IDE	Windows, macOS, or Linux

For this book, installation of Java Development Kit 17+ is required. To verify if it is available on your system, execute the following commands:

- Windows Command Prompt: `java -version`
- Linux or macOS system command line: `java -version`

Expected output:

```
openjdk version "17" 2021-09-14
OpenJDK Runtime Environment (build 17+35-2724)
OpenJDK 64-Bit Server VM (build 17+35-2724, mixed mode,
sharing)
```

In case you don't have the JDK installed on your local machine, search for the appropriate instructions for your platform at `https://dev.java/learn/getting-started-with-java/` and find the matching JDK version at `https://jdk.java.net/archive/`.

To download and install Visual Studio Code, visit `https://code.visualstudio.com/download`.

The following page describes and guides the use of the VSCode terminal: `https://code.visualstudio.com/docs/terminal/basics`.

Download the example code files

You can download the example code files for this book from GitHub at `https://github.com/PacktPublishing/Practical-Design-Patterns-for-Java-Developers`. If there's an update to the code, it will be updated in the GitHub repository.

We also have other code bundles from our rich catalog of books and videos available at https://github.com/PacktPublishing/. Check them out!

Download the color images

We also provide a PDF file that has color images of the screenshots and diagrams used in this book. You can download it here: https://packt.link/nSLEf.

Conventions used

There are a number of text conventions used throughout this book.

Code in text: Indicates code words in text, database table names, folder names, filenames, file extensions, pathnames, dummy URLs, user input, and Twitter handles. Here is an example: "Let us examine the generalization process in the Vehicle class development."

A block of code is set as follows:

```
public class Vehicle {
    private boolean moving;
    public void move(){
        this.moving = true;
        System.out.println("moving...");
    }
```

When we wish to draw your attention to a particular part of a code block, the relevant lines or items are set in bold:

```
sealed interface Engine permits ElectricEngine,
    PetrolEngine  {
    void run();
    void tank();
}
```

Any command-line input or output is written as follows:

```
$ mkdir main
$ cd main
```

Bold: Indicates a new term, an important word, or words that you see onscreen. For instance, words in menus or dialog boxes appear in **bold**. Here is an example: "The bytecode is running a **Java virtual machine (JVM)**."

> **Tips or important notes**
> Appear like this.

Get in touch

Feedback from our readers is always welcome.

General feedback: If you have questions about any aspect of this book, email us at customercare@ packtpub.com and mention the book title in the subject of your message.

Errata: Although we have taken every care to ensure the accuracy of our content, mistakes do happen. If you have found a mistake in this book, we would be grateful if you would report this to us. Please visit www.packtpub.com/support/errata and fill in the form.

Piracy: If you come across any illegal copies of our works in any form on the internet, we would be grateful if you would provide us with the location address or website name. Please contact us at copyright@packt.com with a link to the material.

If you are interested in becoming an author: If there is a topic that you have expertise in and you are interested in either writing or contributing to a book, please visit authors.packtpub.com.

Share Your Thoughts

Once you've read *Practical Design Patterns for Java Developers*, we'd love to hear your thoughts! Scan the QR code below to go straight to the Amazon review page for this book and share your feedback.

https://packt.link/r/1-804-61467-X

Your review is important to us and the tech community and will help us make sure we're delivering excellent quality content.

Download a free PDF copy of this book

Thanks for purchasing this book!

Do you like to read on the go but are unable to carry your print books everywhere? Is your eBook purchase not compatible with the device of your choice?

Don't worry, now with every Packt book you get a DRM-free PDF version of that book at no cost.

Read anywhere, any place, on any device. Search, copy, and paste code from your favorite technical books directly into your application.

The perks don't stop there, you can get exclusive access to discounts, newsletters, and great free content in your inbox daily

Follow these simple steps to get the benefits:

1. Scan the QR code or visit the link below

https://packt.link/free-ebook/9781804614679

2. Submit your proof of purchase
3. That's it! We'll send your free PDF and other benefits to your email directly

Part 1: Design Patterns and Java Platform Functionalities

This part covers the purpose of software design patterns. It outlines the fundamental ideas of the object-oriented programming APIE and SOLID design principles and provides an introduction to the Java platform, which is crucial in understanding how to effectively utilize design patterns.

This part contains the following chapters:

- *Chapter 1, Getting into Software Design Patterns*
- *Chapter 2, Discovering the Java Platform for Design Patterns*

1

Getting into Software Design Patterns

Every software architect or developer often faces the challenges of structuring code – how to develop a code structure that remains sustainable, just as an artist draws their painting. This chapter will take us on a journey into writing program code. You will explore the challenges behind the structure of code and its organization. Together, we will approach the topic from an early stage described by the pillars of object-oriented programming, known as APIE. We will also review the principles of SOLID to gain clarity in understanding design patterns.

In this chapter, we will cover the following topics:

- Code – from symbols to program
- Examining OOP and APIE
- Understanding the SOLID design principles
- The significance of design patterns
- Reviewing what challenges design patterns solve

By the end of this chapter, you will have reviewed the basic programming concepts, which will form the basis of the rest of the book.

Technical requirements

You can find the code files for this chapter on GitHub at `https://github.com/PacktPublishing/Practical-Design-Patterns-for-Java-Developers/tree/main/Chapter01`.

Code – from symbols to program

Human speech is fruitful, rich, colorful, and way beyond what the words themselves may express. Nouns, verbs, and adjectives for precisely expressing a moment or action can be used. In contrast, machines do not understand the complex constructions or expressions that humans are able to create.

Machine language is limited, well-defined, extremely specific, and simplified. Its goal is to provide the precise expression of intent for which it is designed. This contrasts with human language whose purpose is just communication and not necessarily with specifics.

A machine's intent can be expressed as a defined instruction or a set of them. This means that machines understand the instructions. These instructions must be available to the machine in some form at the time of execution. Each machine normally has a set of instructions. Based on this kind of instruction set, machines can perform the required instructions, as shown here:

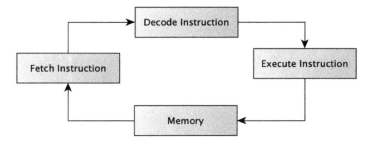

Figure 1.1 – A simplified instruction cycle inside the CPU (instruction
is taken from memory and the result is stored)

Let us explore one individual instruction. The instruction can be understood as a command given to the processor. The processor is the heart of the machine, or the center of the ordering and executing of processes. The machine may contain one or more of them. It depends on its design, but in any case, there is always one that takes the lead. For further simplification, we will only consider one – that is, consider a system that only has one **central processing unit** (**CPU**) dedicated to executing a program.

A CPU is a device that executes instructions containing a computer program. The CPU must contain such an instruction set, as shown in the previous diagram, to process the requested action.

Because instructions can take completely different forms depending on the CPU, there is no defined standard. This promotes different CPU platforms, which is not necessarily a bad thing and contributes to evolution. However, the fact remains that the instructions are not easy for people to read.

We have stated that machines can perform instruction collection, ideally as a continuous flow. The flow of instructions can be simplified as a queue in memory, where one instruction goes in and the other leaves. The CPU plays the role of an interpreter who works with this memory cyclically (as we saw in *Figure 1.1*). Okay, so the CPU interprets, but as the instructions are added to the memory, where do they come from, and how can such a stream be created?

Let us gather some thoughts. Machine instructions, in most cases, originate from a compiler.

What is a compiler? The compiler can be viewed as a CPU or a platform-specific program that translates text into target actions. The text we use to call the program and the result could be named machine code. The following diagram illustrates this:

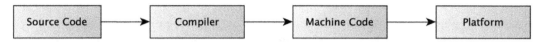

Figure 1.2 – A simplified platform-specific flow from the source code
through the compiler program to its resultant action

Machine code is a low-level language that the machine understands and consists of language instructions that are processed sequentially (see *Figure 1.1*); the program was compiled, executed, and run.

In the case of Java, there is no machine code:

Figure 1.3 – A simplified flow for the Java program through the compiler to its platform execution

The source code is compiled by the Java compiler into bytecode. The bytecode is running a **Java virtual machine (JVM)** (see *Figure 1.3*). In this situation, the JVM plays the role of the interface between the bytecode and the actual instructions that are executed on the CPU. The JVM emulates a bytecode instruction. It does this using the **just-in-time (JIT)** compiler that is part of the JVM. The JIT compiler translates bytecode instructions into native processor instructions. The JVM is a platform-specific interpreter, analogous to directly compiled code (see *Figure 1.2*). The JVM also provides additional features such as memory management and garbage collection, which is what makes the Java platform so powerful. All these features allow developers to write code once, compile it into bytecode, and run a supported platform – known as **write once, run anywhere (WORA)**.

In the context of the previous exploration, Java is a high-level language that is translated to a low level. Java provides a strong abstraction from the details of computer functionality. It allows programmers to create simpler programs for complex challenges.

At this point, we begin our journey of jointly exploring standardized solutions. Later in the book, we will review how to create code that is maintainable and extensible with fewer memory requirements. Together, we will discuss different types of design patterns that can help us to make our daily work understandable, transparent, and more fun.

Examining OOP and APIE

In the previous section, we learned how a program written in one of the high-level languages is converted into machine instructions that are processed by the CPU. The high-level language provides a framework for expressing the desired ideas by following the details of the language implementation. Such languages commonly provide many neat constructions or statements that do not limit the imagination. In **object-oriented programming** (**OOP**) language, the representation of the core carrier is presented by the concept of the object. This book focuses on the Java language. Java is a fully object-oriented language with additional features. What does object-oriented language mean exactly? In computer science, this means that the program focuses on the concept of classes, where instances of these classes represent an object. Next, we will repeat the importance of the OOP paradigm and deal with some basic concepts.

These terms can be expressed by the abbreviation of **abstraction, polymorphism, inheritance, and encapsulation** (**APIE**). The letters APIE indicate the four basic pillars of OOP languages. Let's examine each word in a separate section in reverse order – so, EIPA. The motivation is to bring more clarity to our understanding of the concept of OOP.

Only exposing what's required – encapsulation

The first in reverse order is encapsulation – let's start with it. OOP languages, including Java, work with the concept of classes. Imagine that a class is a vehicle. The class provides all the fields that can be statically typed or object-specific – that is, initiated after an object is instantiated in the allocated memory. The concept is similar with respect to class or object methods. The method may belong to a class or its instance – in the considered example, to a vehicle. Any method can work over an object or class field and change the internal state of the vehicle or the field values (see *Example 1.1*):

```
public class Vehicle {
    private boolean moving;
    public void move(){
        this.moving = true;
        System.out.println("moving...");
    }

    public void stop(){
        this.moving = false;
        System.out.println("stopped...");
    }
}
```

Example 1.1 – The Vehicle class hides an internal state (moving)

We can apply encapsulation to the example of a vehicle. We imagine a real vehicle – only one. In such an imaginary vehicle, all internal elements and internal functions remain hidden from the driver. It only exposes the functionality it serves, such as the steering wheel, which the driver can control. This is the general principle of encapsulation. The state of an instance can be changed or updated through exposed methods or fields; everything else is hidden from the outside world. It is quite a good practice to use methods to modify the inner array or arrays of an instance. But we will repeat that later in this book. So far, it's just a good hint.

Inevitable evolution – inheritance

In the previous section, an instance of an imaginary vehicle class was created. We encapsulated all the functions that should not be exposed to the driver. This means that the driver may not know how the engine works, only how to use it.

This section is devoted to the property of inheritance, which we will demonstrate in the following example. Assume that the vehicle's engine is broken. How can we replace it? The goal is to replace the current one with a functional one. An engine that works this way may not necessarily be the same, especially if the vehicle model already has old parts that are not available on the market.

What we do is derived from all the attributes and functions needed to create a new engine. Concerning the class, the new replacement module will be a child in the class hierarchy.

Although the engine will not be a perfect replica and does not have the same unique object identifier, it will match all the parent properties.

With that, we have described the second pillar of inheritance in OOP – the ability to create a new class above the existing subclass. However, software designers should be wary of the fourth pillar, encapsulation, and any violations caused by a subclass depending on the implementation details of its superclass.

Behavior on demand – polymorphism

The third concept is polymorphism. With a little imagination, this can be understood as "many forms." So, what does that mean here?

Given the vehicle described previously, it could be defined as the ability to perform a particular action in many ways. This would mean, in the context of a vehicle, that the movement of the other method, move, could happen differently based on the inputs or the state of the instance.

Java allows for two types of polymorphism, both of which differ in their runtime behavior. We will discuss both in detail.

Method overloading

This type is known as static polymorphism. This means that the correct method is resolved during program compilation – so, at compile time. Java provides two types of method overloads:

- Changing the input argument type:

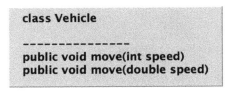

Figure 1.4 – Overloading the method of the Vehicle class by changing the input types

- Changing the number of method arguments:

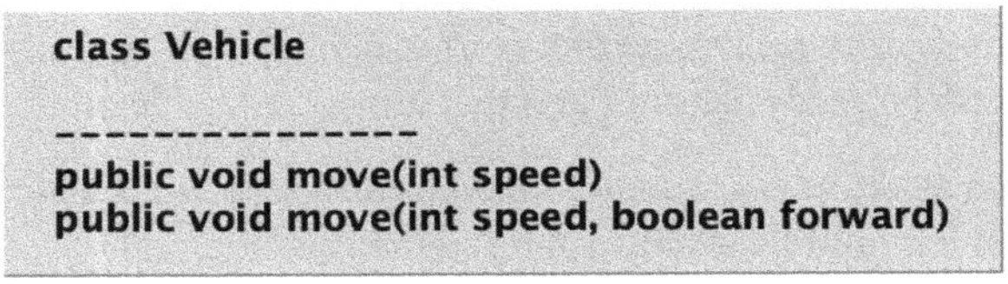

Figure 1.5 – Overloading the method of the Vehicle class by changing the number of arguments

Now, let's look at the second type of polymorphism.

Method overriding

This is sometimes called dynamic polymorphism. This means that the method performed is known at runtime. The overridden method is called through reference to the object instance of belongingness. Let us examine a simple example to illustrate this. Consider the `Vehicle` class a parent class (see *Figure 1.6* and *Example 1.2*) with a method called `move`:

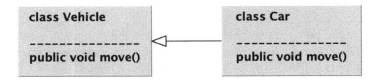

Figure 1.6 – The relation between the overridden move methods for the parent and child classes

We intend to create a child class, `Car`, with a similar method named `move`. The child provides slightly different functions because the `Car` instance moves faster than the parent instance, `Vehicle`:

```java
public class Vehicle {
    public void move(){
        System.out.println("moving...");
    }
}
public class Car extends Vehicle {
    @Override
    public void move(){
        System.out.println("moving faster.");
    }
}

Vehicle vehicle = new Car();
vehicle.move();

output: moving faster...
```

Example 1.2 – The Vehicle variable holds the reference to the Car instance and the appropriate move method is executed at runtime (see Figure 1.6)

We will touch on this topic in more detail in *Chapter 3, Working with Creational Design Patterns*.

Standard features – abstraction

The last letter to cover (but the first letter in the abbreviation APIE) leads us to the hitherto unspecified pillar of abstraction. The key to this concept is the constant removal of specifics or individual details to achieve the generalization of the purpose of the object.

To get the best experience with this concept, let us get into the context with the vehicle example. We do not intend to describe a specific car model that belongs to a group of vehicles. Our goal is to define a common functionality that all types of vehicles under consideration can include in the context of our efforts. With such knowledge, we create a suitable abstraction, an abstract class that can be inherited later when constructing a particular model class (see *Example 1.3*).

This approach allows us to focus our efforts on generalizing and abstracting vehicle characteristics. This can have a positive impact on code reduction and reusability.

The abstraction in Java can be achieved in two ways:

- Abstract classes with abstract methods (see *Example 1.3* and *Figure 1.7*):

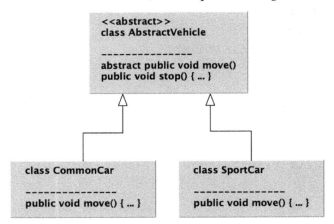

Figure 1.7 – The AbstractVehicle class with its CommonCar realizations and SportCar classes

```java
public abstract class AbstractVehicle {
    abstract public void move();
    public void stop(){
        System.out.println("stopped...");
    }
}
public class CommonCar extends AbstractVehicle{
    @Override
    public void move() {
        System.out.println("move slow...");
    }
}
public class SportCar extends AbstractVehicle{
    @Override
    public void move() {
        System.out.println("move fast...");
    }
}
```

Example 1.3 – The extraction of the common functionality without providing a particular implementation by using an abstract class concept

- Using interfaces (see *Example 1.4* and *Figure 1.8*) with a generic abstract method:

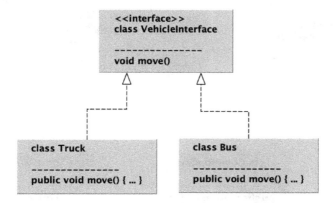

Figure 1.8 – The abstraction concept achieved by using interfaces

```java
public interface VehicleInterface {
    void move();
}
public class Truck implements VehicleInterface{
    @Override
    public void move() {
        System.out.println("truck moves...");
    }
}
public class Bus implements VehicleInterface{
    @Override
    public void move() {
        System.out.println("bus moves...");
    }
}
```

Example 1.4 – A similar functionality extraction by using Java interfaces

Both concepts of abstraction can be combined (see *Figure 1.9*):

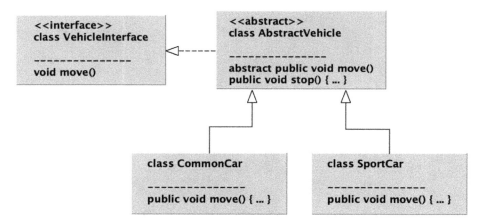

Figure 1.9 – A combination of both abstraction concepts

Abstract classes and interfaces have their place in the design of code structure. Their use depends on demand, but both have a very positive impact on code maintainability and help in the use of design patterns.

Gluing parts to APIE

The motivation for each of the pillars mentioned in the previous sections is to introduce structure into the code through a given set of concepts. The pillars are defined and complementary. Let's just examine one unit, the Vehicle class, and its instance. Instance logic and data are encapsulated and exposed through methods to the outside world. Vehicle characteristics can be inherited so that a new vehicle design, such as a new model, can be specified. Exposed methods can provide model-based behavior and incoming arguments with internal instance state changes. When crystalizing thoughts about a new vehicle, we can always generalize its behavior and extract it using an abstract class or interface.

Let us examine the generalization process over the Vehicle class development. When preparing to define a new vehicle model, we can always generalize its characteristics and extract it using an abstract class or interface. Let's look at the following diagram:

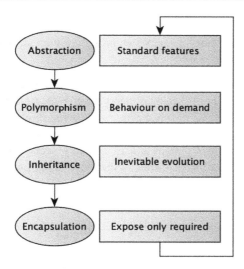

Figure 1.10 – APIE viewed as a continual improvement process

Although these four pillars seem trivial, it is incredibly difficult to follow them, as we will continue to show in the following sections and chapters.

So far in this section, we learned about the four basic pillars of OOP and examined how these principles affect code design. Next, we will learn more about sustainable code design concepts. Let us roll on to the following section.

Understanding the SOLID design principles

In the previous sections, the idea of structured work was introduced. The development pillars of APIE were elaborated on in detail using examples. You have gained a foundational understanding of the concept of class instances in terms of object-oriented principles and how we can create different types of specific objects:

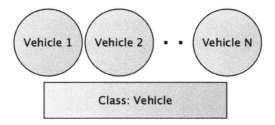

Figure 1.11 – Vehicle N, where N is a positive integer number, represents an instance of the Vehicle class

Classes can be instantiated so that an instance becomes an object. The object must fit into free memory. We say that the object allocates memory space. When Java is considered, allocated memory is virtual space inside the physical system's memory.

Just a small note – we previously discussed the existence of the JVM, an interpreter of compiled bytecode for the required platform (see *Figure 1.3*). We mentioned other JVM features, one of which is memory management. In other words, the JVM assumes responsibility for allocating virtual memory space. This virtual memory space can be used to allocate an instance of a class. This virtual memory and its fragmentation are taken care of by the JVM and an unused object cleans up the selected garbage collection algorithm, but this is beyond the scope of this book and would be the subject of further study (see *Reference 1*).

Every programmer, although it may not be obvious at first glance, plays the role of a software designer. The programmer creates the code by writing it. The code carries an idea that is semantically transformed into action depending on the text entered.

Over time, software development has gone through many phases and many articles have been written and published on software maintenance and reusability. One of the milestones in software development may be considered the year 2000 when Robert C. Martin published his paper on *Design Principles and Design Patterns* (see *Reference 2*). The paper reviews and examines techniques in the design and implementation of software development. These techniques were later simplified in 2004 into the mnemonic acronym SOLID.

The goal of the SOLID principles is to help software designers make software and its structure more sustainable, reusable, and extensible. In the following sections, we will examine each of the individual terms hidden after the initial letter in the abbreviation SOLID.

The single-responsibility principle (SRP) – the engine is just an engine

The first principle is a well-defined class goal. We can say that each class should have only one reason to exist. As in, it has the intention and responsibility for only one part of the functionality. The class should encapsulate this part of the program. Let's put this in the context of an example. Imagine the previous example of a vehicle and its abstraction. We are now extending this class with the Engine and VehicleComputer classes, as shown:

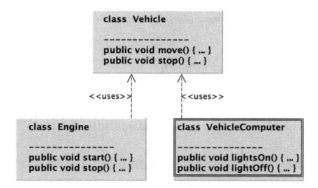

Figure 1.12 – The `Vehicle` class instance using `Engine` and `VehicleComputer`
realization but an engine functionality does not interfere with the lights

The engine can start and stop, but the instance of the `Engine` class cannot control vehicle lights, for example. The light control is the responsibility of the vehicle computer class instance.

The open-closed principle (OCP)

This principle states that the class or entity under consideration should be open to extension but closed to modifications. It goes hand in hand with the concepts already mentioned. Let's put this in the context of an example where we consider the `Car` and `Truck` classes. Both classes inherit the `Vehicle` interface. Both believe that vehicle entities have a move method.

By not thinking about proper abstraction and without respecting the OCP, code can easily bear unexpected difficulties when classes are not easy to reuse or cannot be handled (see *Example 1.5*):

```
public interface Vehicle {}
public class Car implements Vehicle{
    public void move(){}
}
public class Truck implements Vehicle {
    public void move(){}
}
-- usage --
List<Vehicle> vehicles = Arrays.asList(new Truck(), new
    Car());
vehicles.get(0).move() // ERROR, NOT POSISBLE!
```

Example 1.5 – Although both are considered entities, Truck and Car inherit a Vehicle interface, the move method is compliant, and this causes an issue in extension or execution

The correction of the example at hand is very trivial in this case (see *Example 1.6*):

```
public interface Vehicle {
    void move();      // CORRECTION!
}
--- usage ---
List<Vehicle> vehicles = Arrays.asList(new Truck(), new
    Car());
vehicles.get(0).move() // CONGRATULATION, ALL WORKS!
```

Example 1.6 – The Vehicle interface provides a move abstraction method

Obviously, as code evolves, non-compliance leads to unexpected challenges.

The Liskov Substitution Principle (LSP) – substitutability of classes

The previous sections dealt with inheritance and abstraction as two of the key pillars of OOP. It will come as no surprise to those of you who have read carefully that, given the class hierarchy of parent-child relationships, a child may be replaced or represented by its parent and vice versa (see *Example 1.7*). Let us look at the example of CarWash, where you can wash any vehicle:

```
public interface Vehicle {
    void move();
}
public class CarWash {
    public void wash(Vehicle vehicle){}
}
public class Car implements Vehicle{
    public void move(){}
}
public class SportCar extends Car {}
--- usage ---
CarWash carWash = new CarWash();
carWash.wash(new Car());
carWash.wash(new SportCar());
```

Example 1.7 – A CarWash example where any Vehicle type can be substituted by appropriate instances of classes in the class hierarchy

This means that classes of a similar type can act analogously and replace the original class. This statement was first mentioned during a keynote address by Barbara Liskov in 1988 (see *Reference 3*). The conference focused on data abstraction and hierarchy. The statement was based on the idea of substitutability of class instances and interface segregation. Let's look at interface segregation next.

The interface segregation principle (ISP)

This principle states that no instance of a class should be forced to depend on methods that are not used or in their abstractions. It also provides instructions on how to structure interfaces or abstract classes. In other words, it controls how to divide the intended methods into smaller, more specific entities. The client could use these entities transparently. To point out a malicious implementation, consider Car and Bike as children of the Vehicle interface, which shares all the abstract methods (see *Example 1.8*):

```
public interface Vehicle {
    void setMove(boolean moving);
    boolean engineOn();
    boolean pedalsMove();
}
public class Bike implements Vehicle{
    ...
    public boolean engineOn() {
        throw new IllegalStateException("not supported");
    }
    ...
}
public class Car implements Vehicle {
    ...
    public boolean pedalsMove() {
        throw new IllegalStateException("not supported");
    }
}
--- usage ---
private static void printIsMoving(Vehicle v) {
    if (v instanceof Car) {
        System.out.println(v.engineOn());}
    if(v instanceof Bike)
        {System.out.println(v.pedalsMove());}
}
```

Example 1.8 – Various implementations of inherited method abstraction

Some of you with a keen eye will already notice that such a software design direction negatively involves software flexibility through unnecessary actions that need to be considered (such as exceptions). The remedy is based on compliance with the ISP in a very transparent way. Consider two additional interfaces, HasEngine and HasPedals, with their respective functions (see *Example 1.9*). This step forces the printIsMoving method to overload. The entire code becomes transparent to the client and does not require any special treatment to ensure code stability, with exceptions as an example (as seen in *Example 1.8*):

```java
public interface Vehicle {
    void setMove(boolean moving);
}
public interface HasEngine {
    boolean engineOn();
}
public interface HasPedals {
    boolean pedalsMove();
}
public class Bike implements HasPedals, Vehicle {...}
public class Car implements HasEngine, Vehicle {...}
--- usage ---
private static void printIsMoving(Vehicle v){
    // no access to internal state
}
private static void printIsMoving(Car c) {
    System.out.println(c.engineOn());
}
private static void printIsMoving(Bike b) {
    System.out.println(b.pedalsMove());
}
```

Example 1.9 – The functionality split into smaller units (interfaces) based on the purpose

Two interfaces, HasEngine and HasPedals, are introduced, which enforce method code overload and transparency.

The dependency inversion principle (DIP)

Every programmer, or rather software designer, will face the challenge of hierarchical class composition throughout their careers. The following DIP is a remarkably simple guide on how to approach it.

The principle suggests that a low-level class should not know about high-level classes. In the opposite direction, this means that the high-level classes, the classes that are above, should have no information about the basic classes at lower levels (see *Example 1.10*, with the `SportCar` class):

```
public interface Vehicle {}
public class Car implements Vehicle{}
public class SportCar extends Car {}
public class Truck implements Vehicle {}
public class Bus implements Vehicle {}
public class Garage {
    private List<Vehicle> parkingSpots = new ArrayList<>();
    public void park(Vehicle vehicle){
        parkingSpots.add(vehicle);
    }
}
```

Example 1.10 – The garage implementation depends on vehicle abstraction, not concrete classes in a hierarchy

It also means that the implementation of a particular functionality should not depend on specific classes, but rather on their abstractions (see *Example 1.10*, with the `Garage` class).

Significance of design patterns

The previous sections introduced two complementary approaches to software design – APIE and SOLID concepts. It has begun to crystallize that having code in a transparent form can be beneficial for a variety of reasons, because every programmer often, if not always, faces the challenge of designing a piece of code that extends or modifies existing ones.

One wise man once said, "*The way to Hell is the path of continual technical debt ignorance….*" Anything that slows down or prevents the development of applications can be considered a technical debt. Translated into a programming language, this would mean that even a small part matters, if not now, then later. It also follows that code readability and purpose are crucial to application logic, as it is possible to verify various hypotheses (for example, application operation).

The inability to perform business-oriented application testing can be considered the first sign of incorrect development trends. It may appear to require the use of different mock-up techniques during verification. This approach can easily turn into providing false-positive results. This can usually be caused by the clutter of the code structure, which forces programmers to use mocks.

Although the SOLID and APIE concepts suggest several principles, they still do not guarantee that the project code base will not start to rot. Adherence to these principles makes it difficult, but there is still room because not all concepts provide the required framework for dealing with rot.

There may be long stories of how software can rot over time, but one fact that remains is that there is a cure for avoiding it or letting it go. The cure is covered by an idea called **design patterns**. The idea of a design pattern not only covers the readability of the code base and its purpose but also advances the ability to verify required business hypotheses.

What are the ideas behind defining it to get more clarity? The design pattern idea can be described as a set of reusable coding approaches that solve the most common problems encountered during application development. These approaches are in line with the previously mentioned APIE or SOLID concepts and have an incredibly positive impact on bringing transparency, readability, and testability to the development path. Simply put, the idea of design patterns provides a framework for accessing common challenges in software design.

Reviewing what challenges design patterns solve

Take a deep breath and think about the motivation for writing the program. The program is written in a programming language, in our case, Java, and is a human-readable form to address a specific challenge. Let's look at it from a different perspective.

We can state that writing a program is considered a goal. The goal has its reason defined by known needs or requirements in most cases. Expectations and limitations are defined. When the goal is known, each action is chosen with the aim of achieving it. The goal is evaluated, organized, and placed in the context of the destination, where the destination means a work program addressing the required challenge. Imagine all the difficulties mentioned in the previous sections.

Day after day, a new solution is posed, instead of a transparent solution. Every day, another local success keeps the project afloat, despite everything looking good on the surface.

Currently, most teams follow the SCRUM framework. Imagine a situation where the team follows the SCRUM framework (see *Reference 4*) and application development begins to deviate from the goal. Daily standup meetings run smoothly from time to time: it is mentioned that a fundamental error has been found. A few days later, the bug is successfully fixed with great applause. Interestingly, the frequency of such notifications is growing – more corrections, more applause. But does this really mean that the project is moving towards its goal? Does this mean that the application works? Let's look at the answer.

There is a darker side – the backlog is growing with features and technical debt. Technical debt is not necessarily a terrible thing. Technical debt can stimulate the project and can be especially useful in the concept validation phase. The problem with technical debt occurs when it is not recognized, ignored, and poorly evaluated – even worse when technical debt starts being labeled as new features.

Although the product backlog should be one entity, it begins to consist of two different and unfortunately incompatible parts – the business and the sprint backlog (mostly technical debt). Of course, the team is working on a sprint backlog that comes from planning meetings, but with increasing technical debt, there is less and less room for the relevant business functions of the product. The trends observed in this way can result in extremely tricky situations during each new sprint planning session, where the development resources should be allocated. Let's stop for a moment and recall this situation where the team cannot move the product forward due to technical debt.

The values of the SCRUM methodology can be simplified to courage, concentration, determination, respect, and openness. These values are not specific to the SCRUM framework. Because the team's motivation is to deliver the product, they all sound very logical and fair.

We will now refresh our memory of the state the team has achieved. A state where it cannot move the project forward and struggles with the definition and proper consolidation of technical departments. This means that the team is doing its job, but may deviate from achieving its ultimate goal. Every discussion is extremely difficult because it is difficult to solve and describe the problem correctly for many different reasons. It may seem that developers may lose their language of communication and begin to misunderstand each other. We can see that the entropy of the software has increased because the coherence is not maintained. The project is beginning to rot and convergence to the inevitable wasted development time increases.

Let us take another deep breath and think together about how to prevent such a situation. It must be possible to identify these tendencies. Usually, each team has some commonality: the team is not always homogeneous in terms of knowledge, but this should not prevent us from identifying the degradation of the learning curve.

The project learning curve can help us identify a rotting project. Instead of gradual improvements towards the goal, the team experiences local successes full of technical repairs and solutions. Such successes do not even correspond to the values of SCRUM and gradual improvement seems unlikely. The solution may not be considered an improvement because it is specific to a particular movement and may violate the specifications of the technology used. During the solution period, the team may not acquire any useful knowledge applicable to the future. This can soon be considered a missing business opportunity due to the inability to supply business elements or only parts of them.

In addition to the degradation of the learning curve, other symptoms can be identified. This can be described as an inability to test a business function. Project code is proving sticky, dependencies are out of control, which can also harm code readability, testability, and, of course, programmer discipline. The daily goal of the software designer can be reduced to closing a ticket.

To avoid getting to this state, this book will provide some guidelines for solving the most common problems in the following chapters by introducing and questioning different types of design patterns. The design patterns are in line with the aforementioned basic pillars of OOP and APIE and promote the principles of SOLID.

What's more, design patterns can highlight any misunderstood directions and enforce the **don't repeat yourself** (**DRY**) principle. As a result, there is much less duplication, code testability, and more fun on the project.

That brings us to the end of this chapter.

Summary

Before we embark on the journey of researching design patterns, let us quickly summarize. This chapter has expanded or improved our understanding of various areas. Each of these areas affects program code from different perspectives:

- Code transparency and readability
- The ability to solve complex challenges
- Following SOLID and OOP principles
- Code testability (it's possible to verify the purpose of the code)
- Easy to extend and modify
- Supporting continual refactoring
- Code is self-explanatory

The program code is written – well done. The next chapter will take us through a survey of the implementation platform – in our case, the Java platform. We will learn in more detail how and what it means to run a program.

Questions

1. What interprets the Java code to the platform and how?
2. What does the acronym APIE represent?
3. What types of polymorphism does the Java language allow?
4. What principle helps software designers to produce maintainable code?
5. What does the OCP mean?
6. What should be considered about design patterns?

Further reading

- *The Garbage Collection Handbook: The Art of Automatic Memory Management*, Anthony Hosking, J. Eliot B. Moss, and Richard Jones, CRC Press, ISBN-13: 978-1420082791, ISBN-10: 9781420082791, 1996.

- *Design Principles and Design Patterns*, Robert C. Martin, Object Mentor, 2000.

- *Keynote address - data abstraction and hierarchy*, Barbara Liskov, `https://dl.acm.org/doi/10.1145/62139.62141`, 1988.

- The SCRUM framework, `https://www.scrum.org/`, 2022.

2

Discovering the Java Platform for Design Patterns

Many years ago, motivated by the lack of a suitable **Application Programming Interface (API)** design, something extraordinary began to happen. In the early days of using the **World Wide Web (WWW)**, the direction of application development was a bit shrouded in fog. In one direction, there was a strong need in the industry to process a large number of database transactions or develop specific proprietary hardware and software. On the other hand, it was not clear what kind of applications might be needed to move the demand forward and how such an application should be maintained.

In this chapter, we will prepare the ground for understanding the value of design patterns from a memory utilization perspective. We will do so by covering the following topics:

- The rise of Java and brief historical facts
- How the Java platform works under the hood
- Exploring Java memory area allocation and management
- How allocated heap is maintained with garbage collection
- Running the first program on the platform
- The threading nature of the Java platform
- Examining the core Java APIs and their values for software design
- Exploring the importance of the Java Platform Module System
- Discovering new helpful platform enhancements
- Introducing Java concurrency

By the end of the chapter, you will have a good understanding of memory allocation on the Java platform, platform guarantees, core APIs, and more. Together with the content of the previous chapter, these topics will form a well-prepared foundation so that you can start with design patterns with full awareness of their benefits.

Technical requirements

The code files for this chapter are available on GitHub at `https://github.com/PacktPublishing/Practical-Design-Patterns-for-Java-Developers/tree/main/Chapter02`.

Knocking on Java's door

In the early 1990s, a small team at Sun Microsystems was formed in order to discover new horizons. The team started with the consideration of extending the C++ features available in those days. One of the goals was to introduce a new generation of software for a small smart device. The introduction of software reusability was a part of this. Small smart devices such as set-top boxes did not have much memory and had to use their resources wisely. The memory, among other things, such as its complexity, error-prone programs, and probably James Gosling's language extension attempt, later led to the rejection of the C++ idea. Instead of struggling with C++, a new language called **Oak** was created in lieu. Due to the trademark issue, the newly created language Oak was renamed Java.

The first public Java version 1.0a.2, together with HotJava Browser, was announced at the SunWorld conference in 1995 by John Gage, the director of science at Sun Microsystems. He was involved in re-directing the Java language from being a language for small hardware devices to being a platform for WWW applications. In these early days, Java was used as part of a website using a technology known as an applet. Java applets were small sandboxes, defined by the frame with limited access and the capability to execute Java bytecode on the local **Java Virtual Machine (JVM)**. Applets resided on a web browser or as a standalone application; they were a very powerful tool that supported one of the basic Java principles, **Write Once, Run Anywhere (WORA)**. However, due to many issues (such as security and stability), the applet technology was marked for removal (Java SE 17).

The Java platform consists of three main parts (*Figure 2.1*):

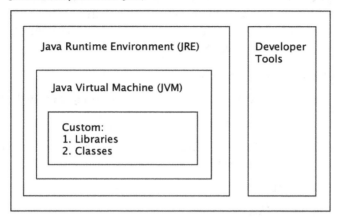

Figure 2.1 – Java Development Kit architecture

These parts are the following:

- A JVM

- The **Java SE (Standard Edition) Runtime Environment (JRE)**

- The **Java SE Development Kit (JDK)**

Let us start an exciting journey through the platform itself and each part.

Exploring the model and functionality of the Java platform

History has shown us that the intended direction can evolve or change: Java is a nice example and is no exception. From its original purpose, it has moved from a platform for smart devices to a platform for entire web solutions, but its development did not stop there. Over the years, Java has become one of the most widely used languages for application development. This can be taken as a side effect of basic hardware independence. It dramatically developed an available set of tools and received a very positive response from a vibrant community.

Let us review each part of the platform (from *Figure 2.1*) individually as it will boost our understanding of writing code.

The JDK

The JDK is a software development environment that provides the tools and libraries needed to develop and analyze Java applications. The JDK provides a collection of basic libraries, functions, and programs needed to compile written code into bytecode. The JDK contains the JRE required to run the application. The JDK also provides some very useful tools, such as the following examples:

- jlink: This helps generate a custom JRE
- jshell: This is a handy **Read-Evaluate-Print-Loop** (**REPL**) tool to try the Java language
- jcmd: This is a utility to send a diagnostic command to the active JVM
- javac: This is the Java compiler, which reads an input file with the .java suffix and produces a Java class file with the .class suffix
- java: This executes a JRE
- Others: Located in the JDK bin directory

The code is written (*Example 2.1*) and stored in a .java file and compiled using the javac command:

```
public class Program {
    public static void main(String... args){
        System.out.println("Hello Program!");
    }
}
```

Example 2.1 – Simple Java program as an executable class that can also be run directly without a compilation step since Java SE 11 (*Reference 26*)

Next, it is possible to create and compile a class with bytecode inside (*Example 2.2*). Run the file using the java command to run the JRE:

```
...
public static void main(java.lang.String...);
descriptor: ([Ljava/lang/String;)V
flags: (0x0089) ACC_PUBLIC, ACC_STATIC, ACC_VARARGS
Code:
stack=2, locals=1, args_size=1
        0: getstatic     #7      // Field java/lang/
            System.out:Ljava/io/PrintStream;
```

```
     3: ldc              #13          // String Hello Program!
...
```

Example 2.2 – Bytecode example from a compiled program displayed by the Java program

The JRE

The JRE is part of the JDK, or it can be distributed as a standalone program for the target operating system. To run a file with a `.class` extension or a **Java Archive** (**JAR**) file, the target system is required to contain the appropriate version of the JRE. Unlike the JDK, the JRE only contains a minimal collection of components needed to run the program, such as the following:

- Core libraries and property files: for example, `rt.jar` and `charset.jar`
- Java extension files: Additional libraries that may reside in the `lib` folder
- Security-related files: Certificates, policies, and so on
- Font files
- Operating system-specific tools

The JRE includes a JVM and precisely two types of compilers:

- **Client Compiler**: Fast loading without optimization. It is designed to run the instructions to obtain a result very quickly. Commonly used for standalone programs.
- **Server Compiler**: Loaded code goes through additional checks to ensure code stability. There is also an effort to produce highly optimized machine code to deliver better performance. It supports better statistics in order to run machine code optimization executed by the **Just-in-Time** (**JIT**) compiler (*Figure 2.2*).

The JVM

Both the JDK and the JRE contain the JVM (*Reference 6*). The JVM is platform-dependent. This means that every system platform requires the use of a dedicated version. Fine, but what does the JVM really do, and how?

Although there are multiple versions of the JVM, even from multiple vendors, the JVM itself is defined by a specification that must be followed. The reference implementation is represented by OpenJDK. In fact, OpenJDK is a collection of several smaller open source projects that may even have different development dynamics, but the OpenJDK release contains planned versions of each.

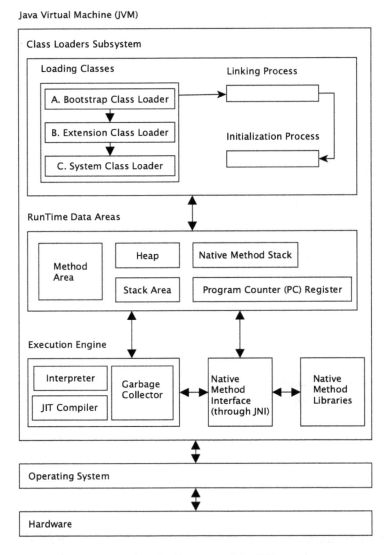

Figure 2.2 – Key parts of the JVM

The OpenJDK JVM implementation (*Figure 2.2*) includes a JIT compiler called **HotSpot** (*Reference 7*). HotSpot is part of the JVM and its responsibility is runtime compilation. In other words, the JIT compiler translates or compiles the provided bytecode into a native system instruction at runtime. This process is sometimes called **dynamic translation**. Due to these JVM dynamic translation capabilities, Java applications are sometimes referred to as system platform-independent and the WORA acronym is used. This statement needs to be abstracted slightly because a JVM system implementation is required to translate the bytecode into a native instruction.

In addition to the JVM JIT compiler, it includes a garbage collector with various algorithms, a class loader, a Java memory implementation model, and a **Java Native Interface** (**JNI**) with libraries (as shown in *Figure 2.2*).

Every JVM provider must follow the specifications. This guarantees that the bytecode will not only be created accordingly but also executed and correctly converted into machine instructions. This means that different vendors may provide different JVM implementations with slightly different metrics or optimizations, such as garbage collector dynamics. These vendors include IBM, Azul, Oracle, and so on. The diversity of vendors can be considered one of the main moving factors for the Java platform's evolution. New features are extended or modified through the **JDK Enhancement Proposal** (**JEP**), where each vendor can contribute or get a very detailed overview.

To summarize, the JVM's responsibilities to remember are as follows:

- Loading linking
- Initiating classes and interfaces
- Program instruction execution

The JVM defines several different areas used by each program (*Example 2.2*). Let's look at each of them one by one, area by area (*Figure 2.2*). This can boost our understanding of the value of design patterns and their approaches, such as the Builder or Singleton pattern.

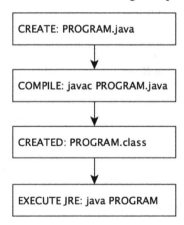

Figure 2.3 – Simplified schema of program compilation and execution

It all starts with written text, representing a program stored in a .java file. The file will be compiled (*Figures 2.3*) and run (*Figure 2.4*) and threads are started. Startup starts the system process in which the JRE is running and the JVM is running as part of the JRE.

Thread	Thread Group ⌄	Thread Id
C2 CompilerThread2	system	9
C1 CompilerThread0	system	10
Signal Dispatcher	system	4
Service Thread	system	5
Sweeper thread	system	11
JFR Recorder Thread	system	13
C2 CompilerThread0	system	7
Monitor Deflation Thread	system	6
C2 CompilerThread1	system	8
Notification Thread	system	16
Reference Handler	system	2
Finalizer	system	3
JFR Periodic Tasks	main	14
main	main	1
JFR Shutdown Hook	main	15
Common-Cleaner	InnocuousThreadGroup	12

Figure 2.4 – Threads that started behind the scene event for a Program.
java execution example (Java Flight Recorder)

With a general idea of the flow, let's start by loading classes into memory.

The class-loader loader area

The class loader subsystem is located in **Random-Access Memory (RAM)** and is responsible for loading classes into memory. The load step consists of the sub-line steps and the first run of the class at runtime. Linking is the process of preparing a class or interface for a runtime environment, which may include internal dependencies, for example. The platform provides internal functions or customized ones; to manage all these capabilities, the platform provides dedicated class loaders:

- **Bootstrap class loader**: Responsible for loading the default platform classes. It is provided by JVM and loads classes from the BOOTPATH (property).

- **Extension class loader**: Loads the additional libraries from the lib/ext directory, which is a part of the JRE installation.

- **System class loader**: The default application class loader that refers to the main method and runs the classes from the served class or module path.

- **User-defined class loaders**: These are instances of ClassLoader and may be used to define custom classes dynamically loading processes to the JVM. It is possible to use a user-defined class destination. Classes can reside on the network, be encrypted inside files, or be downloaded across the network and generated on the fly.

Class loaders work in sequence. The sequence is represented by a hierarchy. This means that every child must refer to its parents. This automatically defines the search order of the binary classes.

When a class is present in RAM, the Java platform takes action to make the class available to the runtime environment. The Java platform runs several processes behind the scenes to move relevant class data to other areas, such as the stack, heap, and so on. Let's look at the stack area next.

The stack area

The stack area (*Figure 2.2*) is reserved for each thread at runtime. This is a small area for storing method references. When a thread executes a method, one entry for that method is created and moved to the top of the stack. This kind of item is called a stack frame, which has a reference to a field of local variables, a stack of operands, and a constant pool to identify the appropriate method. The stack frame is removed when the method is executed normally – that is, without causing any exceptions. This means that local primitive variables such as `boolean`, `byte`, `short`, `char`, `int`, `long`, `float`, and `double` are also stored here, so they are not visible to the second thread. Each thread can pass a copy, but this does not share the origin.

The heap area

The heap is the allocated memory where all instances of the class and array are located. The heap is allocated at startup and is shared among all JVM-initiated threads. Allocated memory is automatically recovered by the automated management system process, also known as **Garbage Collection (GC)**. A local variable can contain a reference to objects. The referenced object is located in a heap.

The method area

The method area is shared across all JVM-initiated threads. The area is allocated during the JVM startup time. It contains runtime data for each class, such as a constant pool, field and method data, the code for constructors, and methods. Probably the most unfamiliar term mentioned is the constant pool. The constant pool is created during the process of loading the class into the method area. It contains the initial values of string and primitive constants, the names of the reference classes and other data needed to properly execute the loaded class, the constants known at compile time, and field references that must be resolved at runtime.

Program counter

The **Program Counter (PC)** register is another important reserved area in memory. It contains a list of created program counters. A PC record is created at the beginning of each thread and contains the address of the currently executed instruction by a specific thread. The address points back to the method area. The only exception is the native method, which leaves the address undefined.

The native method stack

A native method stack record is initiated for each individual thread. Its function is to provide access to native methods through the JNI. The JNI operates with the underlying system resources. Improper usage may turn into two exceptional states:

- The first exception appears when a thread requires more stack space. In this case, a `StackOverflowError` error is thrown and the program crashes, executed with a state higher than 1.

- The second case represents an attempt to add more entries to the stack. The program results in an `OutOfMemoryError` error. It is caused by an attempt to dynamically expand already fully allocated memory space. The memory is insufficient and it is not possible to allocate a new stack for the newly intended thread.

We have examined all the areas required to load and execute a program and we will get acquainted with the areas where the data is located and how they are interconnected. It is slowly becoming clear that in order to achieve stability and maintainability of the program at runtime, it is necessary to design the software in a way that reflects the potential limitations, as the reserved areas correspond to the individual areas.

Let's take a closer look at how the Java platform provides available memory space for each newly created object.

Reviewing GC and the Java memory model

We mentioned the JIT compiler as part of the JVM earlier (*Figure 2.2*). Just to refresh on the JIT compiler, it is responsible for translating the bytecode into system-specific native instructions. These instructions deal with the basic memory and I/O resources available to the program. To properly organize these instructions, the Java platform requires a set of rules that guarantee the program, called bytecode, which must be translated by the JIT compiler at runtime to the same end. Because the Java platform does not use physical memory directly, but rather virtual and cached views, it is very important that the memory management is transparent. The model must provide the required guarantees and is known as the **Java Memory Model (JMM)**.

The JMM

The JMM describes how threads interact with each other through access to allocated memory – the heap (*Figure 2.2*). The execution of a single-threaded program may seem obvious because the instructions are processed in a certain order without external influence and the thread is in isolation. In the case of a single thread (see the `main` method in *Example 2.2* and the `main` thread in *Figure 2.4*), the run areas are modified each time the instruction is executed; there is no surprise. The situation changes when the program starts multiple threads. The JMM enforces its guarantees of reliable Java program execution. The JMM defines a set of rules for possible instruction order changes and execution restrictions caused by sharing objects in memory between different threads. The fact that the JMM strictly follows these rules forces JIT optimization without fear of code instability (maintaining a consistent state).

The rules can easily be reformulated and each action can be changed as long as the execution of the thread does not violate the program order. Basically, this means that the program remains in a consistent state.

Object locks or releases are governed by the order of the program and each thread shares a corresponding memory view of the modified data. The memory view represents the portion of allocated physical memory represented by the heap because each object created is located inside the heap.

One of the important guarantees of the JMM is known as happens-before. It states that one action always happens before another in order to maintain the order of the program. To better understand this rule, it is necessary to describe how system memory works and briefly introduce the general types of memory and how the CPU fits into the process of reading values and executing machine instructions.

Let us start with the CPU. Each CPU contains its own instruction register. The machine code compiled by the JIT compiler has a reference to an available set of instructions. The CPU contains an internal cache used to store a copy of data from the main RAM. The CPU communicates with the reserved RAM. One CPU can run multiple platform threads (depending on the type of CPU) at the same time. The result of this embodiment modifies the state of the RAM in the thread stack or heap. The dedicated RAM for the running Java application is then copied to the CPU cache and used by the CPU registry (*Figure 2.5*).

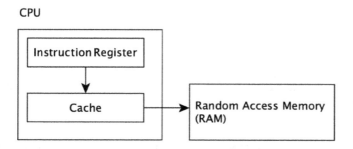

Figure 2.5 – CPU and memory interaction

Those who are attentive may have already noticed that due to memory differentiation, the program can face unpredictable difficulties caused by looking at the program's memory. When multiple threads try to update or read specific values of variables without careful handling, this can result in one of the following problems:

- **A racing condition**: This occurs when two threads attempt to access the same value in an unsynchronized manner.
- **Value update visibility**: A variable update that is shared between multiple threads has not been propagated to the main memory, so other threads get the old value.

To address these challenges, let's analyze a real access to variables. What is already known is that each value is located within the allocated RAM heap. It seems obvious that updating the status of each variable may cause some penalties, as each instruction has to take a whole journey (*Figure 2.5*). In most cases, this is also not necessary. A good example is the implementation of an isolated method (*Example 2.2*). However, there are cases where the actual value of a variable is required from memory, for which the Java platform has introduced the `volatile` keyword. Using the `volatile` keyword before a variable gives the variable a guarantee that when another thread requests a value, it checks its current value in the main memory. This means that using the `volatile` keyword provides a guarantee of happens-before and each thread sees its true value. It is fair to note that because using `volatile` provides a certain level of memory synchronization, it should be used wisely. Its use is associated with performance limitations caused by main memory access.

Another approach to sharing variable values across multiple threads is to use the `synchronized` keyword. Its use gives the method or variable a guarantee that each participant, the thread, will be informed about the approaches. Obviously, the main disadvantage of using `synchronized` is that all threads will be informed about access to the method or variables, which in turn, will cause a decrease in performance due to memory synchronization. As with `volatile`, `synchronized` guarantees happens-before.

The JMM is bright and fresh; we stated that each new object is located in the heap (*Figure 2.2*). We are familiar with the big picture of the JRE architecture, and we know that most Java programs seem to be multi-threaded – there is a set of rules that the Java platform follows so that the process forces the correct order of programs to achieve consistency.

GC and automatic memory management

Although the Java platform may give the impression that the underlying memory is unlimited, this is not true and we will examine it next. So far, we have looked at how variable visibility works across multiple threads and how values are referenced in physical memory. The JMM is just one part of the whole story – let's continue the investigation.

We already know that the Java platform uses an automatic memory management process to maintain the allocated memory for the heap. Part of the process is a program that runs in the background of the daemon thread. It is called the garbage collector (*Reference 5*) and runs silently behind the scenes, reclaiming and compacting unused memory. This is one of the advantages of the dynamic allocation of objects in the heap. Another perhaps less obvious advantage is the ability to work with recursive data structures, such as lists or maps.

GC was invented around 1959 by John McCarthy. The goal was to simplify manual memory management in Lisp. Since then, GC has undergone massive development and various GC techniques have been invented (*Reference 1*). Even after the development of various GC approaches, the security rule remains the most important. The GC should never regain a repository of live objects that contain active references.

Although the developer does not have to bother with memory reclamation, it can be very useful to understand the underlying process to avoid unexpected application failures because system memory is limited in one way or another. The reason remains that even if the GC is in place, it is possible to create code that never loads the object, which means that the application may crash with an `OutOfMemoryError` error.

The goal of the GC process is to keep the heap nice and shiny, ready to allocate a new object. The heap area is divided into smaller segments as shown in *Figure 2.6*:

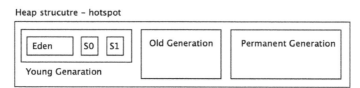

Figure 2.6 – Simplified heap structure divided into promotion segments

It is a known fact that most objects will be allocated and placed in Eden's memory and will not survive the second round of cleanup. After performing Minor GC, all surviving objects are moved to one of the surviving spaces (denoted by **S0** and **S1**). The secondary GC round also checks the **S0** and **S1** fields and spreads them among the others at a time when one of the survival sites may be empty. The object survived; many Minor GCs have been moved to the **Old generation**. The heap also contains a permanent section. The permanent section contains the metadata required by the JVM to describe classes, static methods, and private variables, and is populated at runtime. This area was formerly known as the **Permanent generation** (**Permgen**). It was separated from the main heap memory, was not loaded, and had to be configured. This disadvantage often led to application instability due to insufficient memory requirements. Java SE 8 introduced MetaSpace, which replaced the Permgen concept. MetaSpace has solved the problem of space configuration because it can grow automatically and in addition, garbage can be collected.

GC works essentially in two steps, described as Minor and Major GC. These steps are a proposal for building on the basis of durability – that is, long-term references:

- **Minor GC**: This happens when there is no reference to the object, the object is marked as unreachable, the Young generation area is reclaimed, and the memory can be compacted.

- **Major/Full GC**: An object that has survived several Minor GCs and has been moved to an Old generation heap area. After a long time, it does not refer to any other object, no other object refers to it, and it is ready to be deleted. Full GC is less common than Minor GC and there is a long pause (stop-the-world).

The GC process can be simplified with three steps:

1. In the first step, GC marks unreachable objects (*Figure 2.7*):

    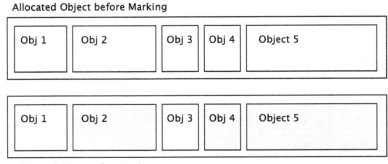

Figure 2.7 – The first collection marking step identifies unused objects in the heap

2. In the second step, the links are removed and the space is left free as it was (*Figure 2.8*):

    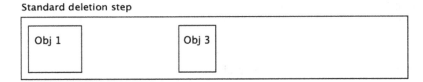

Figure 2.8 – Freeing memory by deleting marked objects

3. The third step is called compacting (*Figure 2.9*). It reorganizes the memory into larger parts, so when a program tries to allocate a larger object, space is ready for it. This makes memory allocation for all objects much faster not only due to the free space but also eliminates the need to scan a free memory frame:

    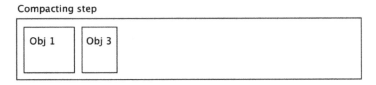

Figure 2.9 – Compacting freed memory to be able to allocate bigger objects in frames

With a fresh awareness of the tasks of the JMM and GC, there is another important concept related to both. The concept of reference types is a way to tell the platform how to handle a specific allocated part of the heap space – more specifically, how to help the platform's internal analytical processes. Reference types have been added to help GC to evaluate the purpose of variables. This means speeding up the decision of whether to collect the variable or not. The concept of a reference type is a neat tool, along with a design pattern, and previously renewed themes make even more sense. The goal of every program is to run as fast as possible. This means that even the waste collection process causes pauses, so it must be as fast as possible. Hence, internal platform processes must also be as fast as possible. So, regardless of the GC algorithm used, when the dataset is small, the process will be much faster. There's also a place for reference types to help keep the allocated memory fresh and clean. The platform offers the following types, sorted by their resilience to GC:

- **Strong references**: The most common type of reference – not required to be specified.
- **Weak references**: References need to be specified manually – `var obj = new WeakReference<Object>();` – and it's a signal to the GC algorithm to reclaim memory during the next GC cycle. This is mostly used during a program initiation phase or caching.
- **Soft references**: These references are reclaimed only when the application is running out of memory. As long as there is no critical need for space, the object stays. The Java platform guarantees that all soft references are cleared before the `OutOfMemoryError` error.
- **Phantom reference**: This represents the weakest type of reference. This team is collected as soon as possible, which means there is no further analysis or promotion to another level. A variable of this type is reclaimed immediately when the GC cycle runs.

Before embarking on the Java API journey, let's quickly summarize our newly acquired knowledge.

References play an important role in the GC process. They tell garbage collectors how to handle a particular variable. The Java memory model provides the required guarantees as to how the value of a variable is read, updated, or deleted. We examined how values are stored in allocated memory, memory segmentation, and their relationship to the underlying system. All this new information helps us with better software design and API usage.

Examining the core Java APIs

The JDK provides a set of tools for creating, compiling, and running the required Java program. We learned how this program uses basic resources to provide the desired result. We have also examined a number of limitations that we must take into account when designing this kind of program. The JDK provides tools for software designers by making an internal collection of classes grouped into APIs available. The previous section explored how the JDK can be extended with external APIs that can be added on demand (discussed earlier in *The JRE* section).

In this section, we will discuss the most important basic APIs we use for design patterns in detail.

Java is an object-oriented language with many other features and extensions. The official basic Java API can be found in the `java.   *` package (as listed in *Table 2.1*).

Sub-package	Description
`java.io.*`	Related to system I/O through data streams, serialization, and filesystems
`java.lang.*`	Automatically imported fundamental classes for the Java language
`java.math.*`	Classes related to the arbitrary precision arithmetic for integers (`BigInteger`) and decimals (`BigDecimal`)
`java.net.*`	APIs related to network protocols and communication
`java.nio.*`	An overview of buffer definitions as data containers and other non-blocking packages
`java.security.*`	Classes and interfaces for the Java security framework
`java.text.*`	Provides classes for handling formatted messages with texts, numbers, and dates
`java.time.*`	APIs for calendars, dates, times, instants, and durations
`java.util.*`	Serves as a collection framework, string parsing, scanning classes, random number generator, Base64 encoders and decoders, and some miscellaneous utilities, among other things

Table 2.1 – java.* packages available in Java 17 SE

Every newly created class can automatically access public classes and interfaces that reside in the `java.lang.*` package from the `java.base` module. As everything is an object, it implies that each class has an `Object` instance.

Primitive data types and wrappers

Java also provides a set of primitive types (*Reference 4*) called literals (*Table 2.2*). One difference between a literal and an `Object` instance is that each literal has a well-defined size in memory. In contrast, the size of the `Object` instance may vary depending on demand. The literal type of Java is signed, which is quite useful to remember if you are dealing with data buffering operations.

Size	Literal name	Range
1 bit (*)	`boolean`	`true` or `false`
1 byte	`byte`	-128 to 127
2 bytes	`short`	-32,768 to 32,767
2 bytes	`char`	\u0000 to \uffff
4 bytes	`int`	-2^{31} to $2^{31}-1$
4 bytes	`float`	-3.4e38 to 3.4e38

Size	Literal name	Range
8 bytes	long	-2^63 to 2^63-1
8 bytes	double	-1.7e308 to 1.7e308

Table 2.2 – Primitive types with their sizes; (*) boolean size is not precisely defined

The primitive types reside in the stack area (refer to *Figure 2.2*) and each literal contains a wrapper object (*Figure 2.10*).

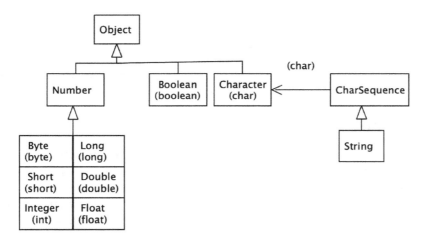

Figure 2.10 – Data type wrapper inheritance with the literals with String type associations

The envelope is initiated around literal values. This means that the literal is stored in the stack area and the wrapper object is located inside the heap. The cover provides additional features. Take the Integer class as an example, which provides the following methods: byteValue, doubleValue, and toString. These methods can be called within a specific design pattern to achieve the desired goal and avoid unnecessary memory contamination. This is in comparison to literals that only provide a native implementation of a value.

The Java platform automatically addresses the literal to the appropriate wrapper class and the like. This fact not only has a bright side but also has a dark side, known as an autoboxing issue (*Example 2.3*). This happens exactly when the primitive type is cast to a wrapper type. This can lead to very frequent waste collection, which can mean an enormous number of stop-the-world events:

```
Int valueIntLiteral = 42;
Integer valueIntWrapper = valueIntLiteral;
```

Example 2.3 – Auto-casting example where a new Integer wrapper is created under the hood

When working with literal numbers, it is useful to keep in mind that a literal with a smaller byte size (*Table 2.2*) can be automatically assigned to a literal with a larger size (*Example 2.4*). The other way around, it causes a compilation error due to the precisely allocated byte size in the memory stack area:

```
byte byteNumber = 1;
short shortNumber = byteNumber;
int intNumber = shortNumber;
```

Example 2.4 – Literal automatic casting

We have checked the numbering and how automatic submission works on the Java platform. A Boolean literal is `true` or `false` and is represented in memory as 1 bit.

The last of the specific literals not yet mentioned is `char` and its cover character. Let's take a closer look because it's also related to the essential `String` object.

Working with the String API

The `String` instance is not literal. A string is represented as an object in Java. It is defined by a sequence of characters. It is almost impossible to avoid using a string to write any program. In addition to the fact that a Java executable requires a `String` field as input to the `main` method, variable names are also represented as a string. The string is immutable in Java. This means that any operation, such as concatenation, on its value will create a new string. More precisely, it is not possible to change its current value. A string is the base class of the Java platform.

A common way to store a string value is to use a String Pool. The String Pool only stores intrinsic values (*Figure 2.11*). This means that it is only possible to have one different constant value present. This approach makes the pool more efficient in terms of memory, including time-consuming string operations.

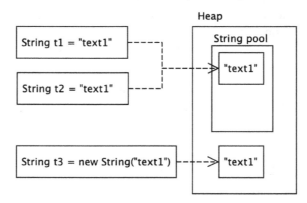

Figure 2.11 – The String Pool is a part of the heap memory and a
String object resides in the heap like other objects

You can also store strings directly in the heap using the `String` constructor – for example, `new String`. In this case, the link is not to an equivalent value that is already present in the String Pool (represented by `t3` in *Example 2.5*) because it is located in a different heap memory space. If you force a search in pool strings, you can use the `intern` method (`t4` in *Example 2.5*):

```
String t1="text1";
String t2="text1";
String t3= new String("text1");
String t4 = t3.intern();
```

Output:

```
1 == t2 => true
t1 == t3 => false
t3 == t4 => false
t1 == t4 => true
```

Example 2.5 – Comparing different ways of assigning String values

Using the + operator on String classes can turn into a very inefficient use of concatenation or program maintainability. To prevent String contamination, the Java platform provides the `StringBuilder` class as part of its APIs. `StringBuilder` prevents temporary values from being stored and only stores the result created by executing its internal `toString` method, which creates a new `String` object in the heap space (*Example 2.6*). `StringBuilder` also introduces the implementation and use of the creational design pattern within the Java SDK:

```
String t5 = new StringBuilder()
          .append("value")
          .append(42)
          .toString();
String t6 = "value42";
```

Output:

```
t5 == t6 => false
```

Example 2.6 – StringBuilder creates a new String object by default in the heap space

We found out how String objects are created and in which heap memory they are stored. This newly acquired information can strengthen us in making decisions by choosing a suitable design pattern or a combination of them to avoid misuse of memory. Because the string is under the hood of an array of characters, primitive type `char []`, the array is not primitive – in fact, it is an object. Let's examine this concept a little more closely because it is also essential for the Java language and platform.

Introducing arrays

To understand the Java collections framework better, first, we will look at an important concept, arrays. In Java, an array is represented by a sequence of the same type of positional index elements. Fields are index-based. Any attempt at runtime to get an element from a non-existent position results in `ArrayIndexOutOfBoundsException`. The array field is allocated as an object and stored in heap space. This means that in the case of insufficient space, an `OutOfMemoryError` exception is thrown. Each array requires a defined size due to memory allocation. Simple field allocation with literals is relatively memory-efficient (*Example 2.8*):

```
int []       array1;
byte[][]     array2;


Object []    array3;
Collection<?>[] array4;
```

Example 2.8 – Multiple array allocation approaches

Arrays allow us to store elements that implement interface classes or a range of abstract classes. The field variable declaration does not create or assign a new field; the variable contains a field reference (*Example 2.9*):

```
int [] a1 = {1,2,3,4};
a1[0] = a1.length;
int e1 = a1[0];


a1.length == 4 => TRUE
a1 instanceof Object => TRUE
```

Example 2.9 – Array initiation, assignment, and verification

The use of the field is often neglected due to its potentially precise requirements and limited auxiliary methods. However, it can help enforce the open-closed principle, which assumes code maintainability.

The field is more often replaced by collection or map structures, which provide additional helper methods. Let us explore the topic more closely.

Discovering a collection framework

Unlike fields, advanced collections provide an automatic resizing feature. This means that the required base representation will be copied and the previous version will become eligible for GC. The Java collections framework includes `List` (*Table 2.3*), `Set` (*Table 2.4*), `Queue` (*Table 2.5*), and `Map` interfaces with several implementations (*Figure 2.12*). Implementations may vary by vendor, but all must conform to basic specifications.

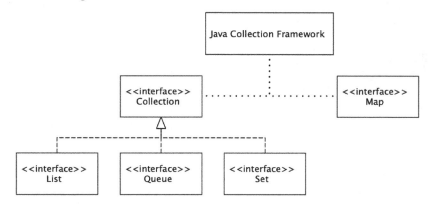

Figure 2.12 – Dependencies between Java collections framework interfaces

The implications of the collection are located in the `java.base` module and its `java.util` package. The package contains the most common implementation, with the known behavior of time complexity. Space complexity is not very relevant, as the framework comes with automatic resizing features. Time complexity can play a more important role in selection when it comes to design patterns, as this can significantly penalize the response of the proposed program. To assess the time complexity of O-notation, O-notation is used to highlight the upper limit and the worst-case program must be used to obtain it.

To evaluate the impact of time complexity, we can go through some nice examples, such as the importance of choosing the right data structure. Let us start with the list structures (*Table 2.3*), which allow access to each element using an index.

Name	Contains	Add	Get	Remove	Data structure
ArrayList	O(n)	O(1)	O(1)	O(n)	Array
LinkedList	O(n)	O(1)	O(n)	O(1)	Linked List

Table 2.3 – Selected List interface implementations with their
time complexities sorted by the actions offered

An algorithm sometimes requires you to verify the presence of an element in the data structure and add or remove a new one. For these cases, let us look at the implementation of the Set interface (*Table 2.4*).

Name	Contains	Add	Remove	Data structure
HashSet	O(1)	O(1)	O(1)	Hash Table
TreeSet	O(log n)	O(log n)	O(log n)	Red-Black tree

Table 2.4 – Selected Set interface implementation sorted by their offered actions and time complexity

The last interface provided by the collection group is Queue (*Figure 2.12*). This data structure is very useful when you only need to work with the first or last element (*Table 2.5*).

Name	Peak	Offer	Poll	Size	Data structure
PriorityQueue	O(1)	O(log n)	O(log n)	O(1)	Priority Heap
ArrayDequeue	O(1)	O(1)	O(1)	O(1)	Array

Table 2.5 – Selected Queue interface implementations with the actions offered and their time complexities

When it comes to implementing a Map interface, it is important to remember what kind of map implementation type is considered. A map represents the structure of a key-value pair. Both the key and the values are descendants of the Object class. Apart from the fact that no literals can be used in the definition or initiation of the map, the correct implementation of the hashCode and equals object methods is required. This requirement is based on the need to identify the correct bucket to resolve potential map collisions. This kind of collision can lead to unexpected time complexity that deviates from our expectations (*Table 2.6*):

Name	Contains Key	Get by Key	Remove by Key	Data structure
HashMap	O(1)	O(1)	O(1)	Hash table
LinkedHashMap	O(1)	O(1)	O(1)	Hash table, linked list

Table 2.6 – Selected Map interface implementations with time complexities by the actions provided

The collections framework uses a heavily behavioral iterator design pattern to traverse through the considered elements. Those who have a keen eye must have already noticed that none of these functions of a collection framework would be possible without a proper mathematical basis. One of the main reasons for using design patterns is to map or create the right structures used by business logic. Let us take a brief look at some basic math features.

Math APIs

Java reveals basic mathematical functions by providing a static implementation of the final `Math` class. Final means that this class cannot be extended, which includes reluctant changes or replacements of basic functions. The `Math` class (*Example 2.10*) is located in the `java.lang` package, which means that it is directly available without the need to import it:

```
Double sin = Math.sin(90);
double abs = Math.abs(-10);
double sqrt = Math.sqrt(2);
```

Example 2.10 – Using common math functions provided by the Math class

Although the `Math` class uses the `random` method, it only gets a `double` result. The `Random` class is in the `java.util` package and provides more customizable capabilities not only for types but also for the required ranges (*Example 2.11*):

```
Random randomNumberWithRange = new Random();
int upperBound = 10;
int randomIntInRange = randomNumberWithRange.
nextInt(upperBound);
double randomDoubleInRange = randomNumberWithRange.
nextDouble(upperBound);
```

Example 2.11 – Generating a random number in a range (0 – upper bound)

The Java `Math` class is also used here, similar to virtually any required calculation that is beyond the capabilities of standard mathematical operators. Using the `Math` class methods can be helpful when functional programming approaches are being followed.

Functional programming and Java

In the previous chapter, we learned about and demonstrated the key principles (APIE) of **Object-Oriented Programming (OOP)**. In recent decades, the Java platform has evolved with the demands of the business and development community. The platform has responded to this challenge by implementing an API that uses the composition of tree functions to provide the desired result. This is in contrast to the traditional loop approach with a collection of imperative commands. This approach caused the larger code base to meet the desired goal.

From Java SE 8 onward, the platform provides an API for streaming (*Reference 15*). It is in the `java.util.stream` package and has nothing to do with Java data streams represented by input and errors (`System.out`, `System.in`, and `System.err`). The Stream API introduces the ability to apply operations to a sequence of elements. There are two types of intermediate operations that can edit or check data, as well as terminal operations. The terminal operation may provide a single result or void. Intermediate operations can be concatenated, but terminal operations terminate the stream. The sequence of elements is lazily evaluated and can also be performed in parallel. By default, performing a parallel stream uses the common **Fork/Join Framework** executor service. The fork-join model can be considered a parallel design pattern that was formulated in the early 1960s (*Reference 17*).

Although the platform allows you to program functional types, OOP concepts remain, followed by strong type requirements. This provides the Stream API with the security that the original element type remains or must be enforced correctly by an intermediate or terminal operation – otherwise, the platform will cause a compilation error. As a reminder, none of these functions would be possible without the introduction of generic types in Java SE 5. Generics (*Reference 4*) allow us to parameterize a class or interface by a type flag to keep compilation safe (*Reference 2*).

Intermediate or terminal operations are implementations of anonymous functions or functional interfaces. They represent a small block of code, formally called a lambda. Let's explain the concept of lambda a little more closely.

Introducing lambdas and functional interfaces

The lambda concept was introduced to enable element operations. Lambdas basically treat data as a code or function as a method. Lambdas rely on the concept of anonymous classes – that is, a class with only one method that performs an action. Java contains a collection of already implemented functional interfaces or ready-to-use functions. Classes are annotated with the @FunctionalInterface annotation, which is a tag available from Java SE 8. It tells the platform that a particular interface contains only one abstract method that can be used to instantiate anonymous classes, as shown in *Table 2.7*. This also means that the interface may contain some default or static functions that belong to the class.

Name	Input argument	Return type	Abstract method	Description
Supplier<T>	-	T	get	Returns a value of type <T>
Consumer<T>	T	-	accept	Consumes a value of type <T>
Function<T, R>	T	R	apply	Consumes a value of type <T> and applies a transformation with return type <R>
Predicate<T>	T	Boolean	test	Consumes an input of type <T> and returns a Boolean result

Table 2.7 – Basic functional interfaces available in the JDK since Java SE 8

Using functional interfaces in lambda expressions

We have discovered that each lambda expression is lazily loaded, which means the code is evaluated on demand, not at compile time, and may be closed by the terminal operation (*Example 2.12*):

```
List<String> list = Arrays.asList("one", "two",
    "forty_two");
list.forEach(System.out::println);
```

Example 2.12 – Converting elements of the List interface to the stream and applying a terminal operation for each Consumer type instance

We can chain the different intermediate functions together (*Example 2.13*) and close the stream with a terminal operation or pass the stream to another method or object:

```
Predicate<Integer> numberTest = new Predicate<Integer>() {
    @Override
    public boolean test(Integer e) {
        return e > 2;
    }
};
String result = Stream.of(1,2,3, 42)
        //.filter(e -> e > 2) //Anonymous class example
        .filter(numberTest)
        .map(e -> "element" + e)
        .collect(Collectors.joining(","));
System.out.println("result: " + result);
```

Example 2.13 – Advanced composition of named and anonymous functional interfaces

The lambda expression stream API plays an important role in the composition of the code. It can be imagined as a process line into which the input object enters and, thanks to a collection of adjustments, the expected result is returned or the action ends. Since the lambdas are evaluated lazily, this means that the process line has a switch. In other words, the Stream API can be considered one of the most important breakthroughs in syntax.

Getting to grips with the Java Module System

One of the main purposes of using a higher-order programming language such as Java is code reusability. A basic building block of the language is the concept of classes according to the principles of APIE. Java can localize these classes into groups defined by specific package names. The package concept encapsulates a group of classes. Classes can provide different levels of visibility to their internal fields and methods. Java specifies the following levels of visibility: `public`, default, `private`, and `protected`. Keywords are used to reduce visibility across different packages to manage their interactions. The way to share a package across an application domain is to keep it public – that is, visible to everyone.

Java has been using the concept of class paths for many years. The class path is a special place where the Class Loader loads its classes. The loaded classes are then used at runtime (denoted as the **Class Loaders Subsystem** in *Figure 2.2*).

However, this concept does not provide any guarantee for the stored package or class. This concept has been considered bad, fragile, and error-prone for many years. A good example is trying to package a JAR executable that contains different versions of libraries with similar package structures and class names. The class path does not differ and the class can be overwritten by different versions.

The breakthrough came with the release of Java SE 9. JSR-376, formerly the core of the Jigsaw project (*Reference 3*), became a common part of the platform. JSR-376 implements the **Java Platform Module System (JPMS)** (*Example 2.14*):

```
$ java -list-modules
java.base@17
java.compiler@17
java.datatransfer@17
<more>
```

Example 2.14 – Listing the available JDK modules for a specific version

Additionally, the platform has been migrated in accordance with the modules (*Example 2.15*):

```
$ java -describe-module java.logging
java.logging@17
exports java.util.logging
requires java.base mandated
provides jdk.internal.logger.DefaultLoggerFinder with
    sun.util.logging.internal.LoggingProviderImpl
<more>
```

Example 2.15 – Describing a java.logging module. The java.base module is automatically present, as it contains the core platform and language functionalities.

The JMPS provides a strong package encapsulation concept that defines application interactions at the package level (*Example 2.16*). The application can be divided into modules that can only detect APIs or services. The JMPS supports package-level dependency building and increases the maintainability, reliability, and security of the application being developed:

```
Module java.logging {
    exports java.util.logging;
    provides jdk.internal.logger.DefaultLoggerFinder with
        sun.util.logging.internal.LoggingProviderImpl;
}
```

Example 2.16 – Example of the module-info.class descriptor exposing a package for external usage

The use of the JPMS is not mandatory. The Java platform uses the JMPS, but if the application is not ready, unnamed modules can be used. In this case, all packages or classes will belong to this kind of unnamed module. In principle, an unnamed module reads each readable module or class from the class path without reflecting any package-level restrictions required by the JPMS. In this way, compatibility with previously developed applications is achieved and the software designer has no doubts about the malfunction of the code base – that is, the JPMS is disabled.

Although the JPMS has incredible potential for application sustainability, security, and reusability, it is often not used because it creates indirect pressure to properly configure the underlying JPMS and use a design pattern that enforces SOLID principles.

When using the JMPS, the platform ensures that the developed application does not contain any cyclic dependencies. Behind the scenes, the JPMS creates an acyclic module graph (not a class path case).

By creating a module descriptor file, the platform provides a set of directives that can be used to expose certain parts of the module to the outside world.

Let's create a simple example of a module to remove any doubt about the use of the JPMS (*Example 2.17*). Our discussion so far can overcome the initial difficulties:

```
module-example
├── example
│   └── ExampleMain.java
└── module-info.java
```

Example 2.17 – Folder structure of module example developed with OpenJDK 17

We create an appropriate executable class, `ExampleMain.java`, and a module descriptor, `module-info.java` (*Example 2.18*). In this way, we tell the platform to use the JPMS:

```
// file module-info.java
module module.example {
    exports example;
}


// file ExampleMain.java
package example;

public class ExampleMain {
    public static void main(String[] args) {
        System.out.println("Welcome to JMPS!");
    }
}
```

Example 2.18 – Simple module example introduced by the file structure in Example 2.17

The example shows how the project could be separated into modules that contain their own descriptors, `module-info.java` files (*Example 2.17*). This descriptor defines an interaction with other modules through dependencies or exposures of module internals. The JPMS ensures that the restrictions, including visibility, are maintained:

```
$ javac -d ./out ./module-example/module-info.java
    ./module-example/example/ExampleMain.java
$ jar -create -file module-example.jar -C ./out .
$ java —module-path ./module-example.jar —module
    module.example/example.ExampleMain
```

Output:

```
    Welcome to JMPS!
```

```
$ java —module-path ./module-example.jar —describe-module
    module.example
```

Output:

```
module.example
exports example
requires java.base mandated
```

Example 2.19 – Steps to compile Example 2.17 with outputs, together with the module descriptor check (Example 2.18) after the compiled result

The JPMS is a big change to the platform and although it opens a new horizon for software designers by providing the ability to define clarity in the package structure, it is not always well received or understood. This may be due to additional requirements that need to be taken into account when designing the system, which essentially relate to the knowledge of APIE or SOLID principles.

The JPMS together with the Stream API, as well as lambdas, may be considered significant changes addressed by the Java SE 11 release – Java SE 11 being the next **Long-Time Support** (**LST**) release after version 8. Let us dive a bit further into some of the changes from Java SE 11 to the next LST version presented by release 17.

A quick review of Java features from 11 to 17+

This version update presents performance and optimization improvements. In this section, we will examine those that are very useful for the specific use of a design pattern and its structure. This equates to platform enhancements that improve code readability, platform usage, or syntax enhancements.

The local variable syntax for lambda parameters (Java SE 11, JEP-323)

Java has often been criticized for the amount of standard code in the use of a variable; Java SE 10 introduced a new keyword, `var`. The derivation of a local type variable lies behind this keyword. It essentially requires that the value type is taken from the newly created reference instance (*Example 2.20*). Using the stream `boxed` function shows a decorator pattern that wraps the stream value with the desired type:

```
Consumer<Integer> consumer = (var number) -> {
    var result = number + 1;
    System.out.println("result:" + result);
};
IntStream.of(1, 2, 3).boxed().forEach(consumer);
```

Example 2.20 – Use of local type inference in a lambda expression and stream shows the reduction of boilerplate code

Although a lambda already allowed an implicit type definition, for example, the use of annotation was not possible.

Switch expressions (Java SE 14, JEP-361)

Software designers have long complained about several inconsistencies in the use of switch commands, such as a control flow problem. Although this enhancement is fully compatible with all controls, it introduces a new form of switch label, `case  CONSTANT->`. The extension also allows more constants to be used, making the entire switch expression more compact. The last improvement is the ability of the switch expression to return its computed value (*Example 2.21*). This has a very positive impact on the implementation of the design pattern, because, for example, behavior types require a precise control flow (*Reference 8*):

```
Var inputNumber = 42;
String textNumber = switch (inputNumber){
    case 22,42 -> String.valueOf(inputNumber);
    default -> throw new RuntimeException("not allowed");
};
System.out.printf("""
        number:'%s'
        %n""", textNumber);
```

Example 2.21 – Compact switch expression usage with a return control flow with a simple text block

Text blocks (Java SE 15, JEP-378)

Many times, you need to create multiple lines with a specific format. Previous use of multiple escape sequences and characters was not as practical, as it could be unpredictable. Text block extension introduces a literal that allows you to represent a string in a predictable way (see the `System.out.printf` method in *Example 2.21*, as well as *Reference 9*).

Pattern matching for instanceof (Java SE 16, JEP-394)

Previously, it was necessary to retype a value type that had already been verified as positive for its type. This increased the code base and sometimes had a negative effect on the stability of the code, even when designing a pattern. This platform extension eliminates the need for a rear cast and the variable can be used directly with the correct type (*Example 2.22, Reference 10*):

```
Object obj = "text";
if(obj instanceof String s){
```

```
        System.out.println(s.toUpperCase());
    }
```

Example 2.22 – Using instanceof with direct type methods

Records (Java SE 16, JEP-395)

The record class type is very useful because its declaration is very simple and can carry all the data needed for the program's business logic. Records carry immutable data. They provide an already implemented hashCode and equals. This means that the designed software does not have to provide additional code (*Example 2.23*, *Reference 11*):

```
private record Example(int number, String text){
    private String getTogether(){
        return number + text;
    }
}
```

Example 2.23 – New record class type may have a very positive impact on code reduction, as it provides generated methods

Sealed classes (Java SE 17, JEP-409)

These are very elegant enhancements to gain control over classes and interfaces, or class extensions and interface implementations (*Reference 12*). Closed classes give software designers wide access to the superclass without the need to extend it. They overcome the limitations of the widely used package access modifier, which previously required the full implementation of abstract methods. The example shows how to define an open class for an extension, the non-sealed keyword (*Example 2.24*):

```
public sealed interface Vehicle permits Car, Bus {
    void start();

    void stop();
}
public non-sealed class Car extends NormalEngine implements
    Vehicle {
    public String toString(){
        return "Car{running="+ super.running +'}';
```

```
        }
    }
```

Example 2.24 – The implementation of the interface methods is provided by the abstract class, NormalEngine

Sealed classes force control over possible extensions (*Example 2.25*) because they provide software with potential security against unwanted software design changes:

```
Public class Motorbike implements Vehicle{
    public void start() {}
    public void stop() {}
}
```

Here's the compilation output:

```
Motorbike.java:2: error: class is not allowed to extend
sealed class: Vehicle (as it is not listed in its permits
clause)
```

Example 2.25 – Sealed classes enforce control over the enhancements

Sealed classes also present some potential problems because the software designer must decide how the newly created classes will be used, indicating whether class extension is allowed, the keyword is unsealed (*Example 2.25*), or the final keyword (*Example 2.26*) is locked:

```
Public final class Bus extends SlowEngine implements
    Vehicle {}
```

Example 2.26 – It's required to decide the class behavior and the Bus class example is locked for any extension

Although this may seem like a possible disadvantage, it provides greater clarity in software development in terms of maintainability and design patterns. This reduces potential unwanted interface or class errors.

UTF-8 by default (Java SE 18, JEP-400)

For many years, unclear encoding has caused issues. Encoding problems were not easy to detect and appeared unpredictably on different system platforms. This enhancement has unified everything and forced UTF-8 as the default encoding (*Reference 13*).

Pattern matching for switch (Java SE 18, Second Preview, JEP-420)

Improvements to `instanceof` fields (JEP-394) and switch case expressions (JEP-361) have made it even better to compress the code base and remove previously unnecessary if-else constructs by using `instanceof` on a very compact, command-oriented controlled command statement: type-served (*Example 2.27, Reference 14*):

```
Object variable = 42;
String text = switch (variable){
    case Integer i -> "number"+i;
    default -> "text";
};
```

Example 2.27 – Compact switch statement with an implicit type match

After reviewing the most important syntactic improvements, we can safely start to delve deeper into one of the main advantages of the platform. Yes, it is the concurrency framework.

Understanding Java concurrency

At the beginning of this chapter, it was shown that even running a simple program (*Example 2.2* and *Figure 2.3*) will cause multiple concretizations of threads (*Figure 2.4*). This means that a program thread created by executing the `main` method did not create any other thread that belonged to that program. The Java platform is well-known and valuable for its ability to perform concurrent or parallel tasks.

We found out how and where initiated threads store their variables and why synchronizing objects located in the heap can lead to unwanted or unexpected program behavior. In this section, we will look at the possibilities of the main threads using the available CPUs.

The motivation for software designers to consider using any concurrent design patterns may be the growing need for better application responsiveness or throughput.

Although the platform already includes the `Thread` class and the `java.lang` package, Java concurrency features such as **executors** have become available to developers since the release of Java SE 5 and can be found in the `java.util.concurrent` package, which is part of the `java.base` module.

Let us look a bit closer.

From a basic thread to executors

The basic build element of the platform is the thread. A thread is represented by an instance of the `Thread` class. The object initiated by the `new` keyword still does not create a platform thread. The object provides a method named `start` that requires explicit use (*Example 2.28*):

```
public class Multithreaded Program {
    public static void main(String[] args) {
        var t = new Thread(() -> {
            while(true){System.out.println("Welcome
                Thread!");}
        });
        t.setDaemon(true);
        t.start();
    }
}
```

Example 2.28 – A simple program with a daemon thread that ends immediately after the JVM stops

Although it may appear that the platform can create an unrestricted `Thread` instance, this kind of statement is not valid. Each newly created thread instance not only takes up heap space or allocates a stack but is also connected to basic system threads (processing cycles) through Java runtime partitions (*Figure 2.2*). This means that uncontrolled thread startup can cause a system error exception due to unavailable resources, insufficient memory, and so on.

The maximum number of system threads created by the Java platform may vary, as it depends on the hardware as well as the JVM configuration. The Java `Thread` class might consider a wrapper for the `Runnable` interface, and the thread accepts its implementation. The `Runnable` interface is another functional interface and requires the implementation of a run method. Starting with the Java SE 8 `Runnable` interface, the instance can be passed to the executor service as an anonymous function.

The Java platform allows you to run a thread that can even survive the termination of the main program, which in many cases, is a reluctant condition and should be considered wisely, as it may block other core resources or stay running.

It is important to remember that the JVM only terminates when all running threads are daemon ones (*Example 2.28*).

Because each thread newly created by the main program is non-daemon, by default, when the sample program is run without an explicit daemon flag, the JVM remains an active process until the base system destroys it.

It allows you to manage uncontrolled thread creation and gives the software designer control over the program's resources and behavior. Java SE 5 introduced the `ExecutorService` and `ThreadFactory` interfaces, where multiple implementations show using a similarly named creational design pattern factory. The `ThreadFactory` interface contains only one `newThread` method, which returns a `Thread` instance. This method logic can accommodate the creation of a new thread and set the group, thread priority, and daemon flag. It also eliminates the number of new thread calls. `ThreadFactory` can be serviced by `ExecutorService` (*Example 2.30*).

Some of the most used executor static method names are as follows:

- `newSingleThreadExecutor()`
- `newSingleThreadExecutor(ThreadFactory threadFactory)`
- `newCachedThreadPool()`
- `newCachedThreadPool(ThreadFactory threadFactory)`

Java SE 5 came up with a concept for the future, a `Future` interface with a generic type of `<T>`. The `Future` interface can be considered an asynchronous calculation that provides a result.

The Java platform provides two different interfaces that can carry thread logic.

Executing tasks

The Java platform provides a thread concept from the beginning represented by the `Runnable` interface and the `Thread` class (*Example 2.29*):

```
ExecutorService executorService =
    Executors.newSingleThreadExecutor();
var runnable = new Runnable(){
    @Override
    public void run() {
        System.out.println("Welcome Runnable");
    }
};
executorService.execute(runnable);
executorService.execute(() -> System.out.println("Welcome
    Runnable"));
```

Example 2.29 – Different approaches to providing the Runnable interface implementation to the executor service, implementation, and anonymous class

Business requirements, along with community expectations, have created a platform for reactive programming or the ability to perform multiple asynchronous payback tasks. As of Java SE 5, the platform provides a `Callable` interface. The `Callable` interface is considered a functional interface. It contains only one abstract method call with a required return type of `<T>`. Because the computation is uncertain, it can cause an exception that must be handled correctly. The `Callable` implementation can be sent to the executor and the started calculation is packed into a future result.

The `Future` instance is the computational work that the base system performs in the background. The interface provides a `get` method (*Example 2.30*) that can be used to retrieve the result. Using this method pauses the current thread and waits until a result is available. Due to the current thread suspension, this method should be used wisely, as it can cause performance penalties:

```
var futureCallable = executorService.submit(callable);
Future<String> futureCallableAnonymous = executor.submit(()
    -> "Welcome to Future");

System.out.println("""
        futureCallable:'%s',
        futureCallableAnonymous:'%s'
        """.formatted(futureCallable.get(),
            futureCallableAnonymous.get()));
```

Example 2.30 – Different approaches to provide a Callable instance to the executor service as a realization or an anonymous class

This contrasts with the `Runnable` interface because the `Callable` interface provides a `Future` instance as a temporary result. The `Callable` exception handling request is also relevant because it can cause logic to be executed or the worker thread can be interrupted. In this case, it is necessary to transfer this to the interim result represented by the `Future` interface.

Summary

In this chapter, we have built a good knowledge base for understanding the internal Java platform. We learned about the differences between statically allocated arrays or methods compared to object instances. We examined the need for proper data synchronization and how Java memory management works and what guarantees the platform provides. We now understand the importance of heap memory, segmentation, and maintenance. We have also already discovered a few frequently used design patterns, which means that when we start implementing any design pattern or collection, we will be aware of the following:

- How fields or variables are handled by the platform

- The importance of memory management
- Specific program error exit states and the reasons for them
- The core APIs provided by the Java platform
- How to utilize functional programming features
- What new enhancements the Java platform provides to make employing design patterns easier
- How to approach Java concurrency challenges

We have built a solid knowledge base over the first two chapters. We will now begin to present pattern by pattern. The next chapter will take us on a journey through creational design patterns. Creational design patterns intensify our awareness of the code structure and how to create sustainable solutions. Let us roll.

Questions

1. Which elements make up the Java platform?
2. What does statically typed language mean?
3. What are the Java language literals?
4. What is responsible for memory reclamation in the Java memory management concept?
5. What are the collections in the Java collections framework?
6. What kind of elements store `Map`?
7. What is the time complexity of retrieving an element from `Set`?
8. What is the time complexity of verifying an element's existence in `ArrayList`?
9. What functional interface is used in the `filter` method of the Stream API?
10. How are elements evaluated in the Stream API?

Further reading

- *The Garbage Collection Handbook: The Art of Automatic Memory Management*, Anthony Hosking, Eliot B. Moss, and Richard Jones, CRC Press, ISBN-13: 978-1420082791, ISBN-10: 9781420082791, 1996
- Java Generics: `https://docs.oracle.com/javase/tutorial/java/generics/index.html`
- The JPMS (JSR 376): `https://openjdk.java.net/projects/jigsaw/spec/`
- The Java tutorials: `https://docs.oracle.com/javase/tutorial/java`

- Java GC basics: https://www.oracle.com/webfolder/technetwork/tutorials/obe/java/gc01/index.html

- The JVM specification, Java SE 17 Edition: https://docs.oracle.com/javase/specs/jvms/se17/html/index.html

- OpenJDK, HotSpot runtime overview: https://openjdk.java.net/groups/hotspot/docs/RuntimeOverview.html

- JEP 361: Switch Expression: https://openjdk.java.net/jeps/361

- JEP 378: Text Blocks: https://openjdk.java.net/jeps/378

- JEP 394: Pattern matching for instanceof: https://openjdk.java.net/jeps/394

- *JEP 395: Records*: https://openjdk.java.net/jeps/395

- *JEP 409: Sealed Classes*: https://openjdk.java.net/jeps/409

- *JEP 400: UTF-8 by Default*: https://openjdk.java.net/jeps/400

- *JEP 420: Pattern Matching for switch (Second Preview)*: https://openjdk.java.net/jeps/420

- The java.util.stream package: https://docs.oracle.com/javase/8/docs/api/java/util/stream/package-summary.html

- *JEP 300: Launch Single-File Source-Code Programs*: https://openjdk.java.net/jeps/330

- *A multiprocessor system design.* Fall Join Computer Conference, Melvin E. Conway (1963). pp. 139 -146.

Part 2: Implementing Standard Design Patterns Using Java Programming

Design patterns are often classified into three well-known categories: creational, behavioral, and structural. This part will explore and demonstrate design patterns from each of these categories. It will show the types of challenges addressed by each design pattern with practical real-world examples.

This part contains the following chapters:

- *Chapter 3, Working with Creational Design Patterns*
- *Chapter 4, Applying Structural Design Patterns*
- *Chapter 5, Behavioral Design Patterns*

3

Working with Creational Design Patterns

In recent decades, the IT community has experienced a dramatic shift from previously isolated systems to distributed or hybrid solutions. These approaches bring to light new possibilities for software development.

Distribution solutions may appear to meet the migration needs of legacy systems, but the reality may prove otherwise. The required refactoring can cause additional problems due to the division of responsibilities or refactoring of tightly coupled logic and business rules and many unknown, hidden logics that are discovered too late to react to.

In this chapter, we will explore creational design patterns. These patterns play a vital role in the software composition. They are very useful for achieving maintainability or readability of the code base. Creational design patterns attempt to follow all the previously noted principles or the **don't repeat yourself** (**DRY**) approach. Let's dive deeper into specific patterns in the following order:

- Applying the factory method pattern
- Instantiating additional factories in encapsulation with the abstract factory pattern
- Creating a different configuration of object instances with the builder pattern
- Avoiding a repeatedly complex configuration with the prototype pattern
- Examining only one instance presence with the singleton pattern
- Speeding up runtime with prepared objects by using the object pool pattern
- Controlling instances on demand with the lazy initialization pattern
- Reducing the object instances with the dependency injection pattern

By the end of this chapter, you will have built up a solid understanding of how to write maintainable code to create objects that can reside on the JVM's heap or stack.

Technical requirements

You can find the code files for this chapter on GitHub at `https://github.com/PacktPublishing/` `Practical-Design-Patterns-for-Java-Developers/tree/main/Chapter03`.

It all starts with a class that becomes an object

In Java, every object must first be described by a class. Let us briefly introduce a common theoretical scenario of a software application. Such scenarios are often divided into the following parts:

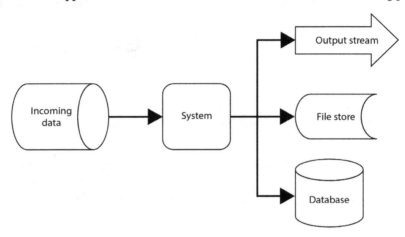

Figure 3.1 – Common application data process from a big picture

The incoming input data stream (that is, the information flow) has been accepted by an application. The application processes the input and creates the result. The result is stored and subjected to the required targeting by a system.

Such a system has the capability to satisfy several different processes under various conditions. The results are stored in several ways, such as a database or a file, or possibly embedded into an intended output stream, such as a web page, to display information to the user.

The system acts as a reservoir of the incoming information flow, processes and stores it in the database, and provides the result. Most of the time, everything is tightly coupled and interconnected.

Coupling has been happening on different levels, without the software designer noticing. The tight coherence was between classes, objects, or even packages. In many ways, it was possible to correct the application performance weaknesses with the more powerful hardware. The system's evolution progressed more or less as a statistical observation of Moore's law, which was published in 1965.

Moore's law stated that each year, the number of components per integrated circuit doubles. The law was revised in 1975 to state that the number of components doubles every *two* years. Although the debate over the law's validity may turn controversial, current trends (and the speed with which hardware upgrades are needed) show that the time for another review is coming. It may not be necessary worldwide to speed up the hardware upgrade (already so fast) because it may not have any effect on the speed of processing information. This observation addresses the feature requirements of a software application, focusing more on the quality and complexity of the implemented algorithms.

It may not be possible to constantly increase the rate at which an object is instantiated due to physical limits, as such information must be physically stored in memory. It means that in the coming decades, we can expect an increase in pressure to improve the efficiency of software and design. To gain clarity of the application logic, it needs to be crystal clear how the application works, and moreover, how the application feeds the key JVM areas, namely, the method stack and heap followed by the thread utilization through the stack areas (as shown previously in *Figure 2.2*).

Due to the current software applications' trends focused on mapping, transforming, or managing a large amount of data, creational design patterns are worthwhile to study, understand, and learn how to deal with common scenarios. Although the time of the **Gang of Four** (**GoF**) book has passed, evolution is inevitable and the challenges remain. In many cases, with proper abstraction, the initial creational design patterns are applicable. Creating objects and class instances, and filling out the intended parts of the JVM, may drastically influence the computation and performance costs, as well as enforce business logic clarity.

In the next section, we discuss different possibilities for object creation. We will also consider the recently added Java syntactic features and possibilities, which should reduce the source code's verbosity. Let us start with one of the most common patterns.

Creating objects based on input with the factory method pattern

The primary purpose of this pattern is to centralize the class's instantiation of a specific type. The pattern leaves the decision to create the exact class type up to the client at runtime. The factory method design pattern was described in the GoF's book.

Motivation

The factory method pattern enforces the separation of code and its responsibility for creating new instances of the class, that is, such a method provides the expected result. The factory hides an application class hierarchy based on a generics abstraction and introduces a common interface. It transparently separates the instantiation logic from the rest of the code. By introducing the common interface, the client gains the freedom to decide on a particular class instance at runtime.

The pattern is often used in the early stages of an application because it is simple to refactor and provides a high level of clarity.

Although this can introduce a bit of complexity, the pattern is easy to follow.

Finding it in the JDK

The factory method pattern is often utilized in the Java Collection framework to construct the desired type. The framework implementations reside in the `java.util` package of the `java.base` module. This package contains different implementations of `Set`, `List`, and `Map`. Although the `Map` type is a valid member of the Java Collection framework, it does not inherit the `Collection` interface as it implements `Map.Entry` to store element tuples, keys, and values. Each implementation of `Set`, `List`, and `Map` provides overloaded `of` factory method to create an instance.

The `Collections` class is a utility class. It contains several factory methods for creating specific collections, such as a list of individual items, a map, or a set. Another useful example of the factory method pattern usage is the `Executors` utility class, which can be found in the `java.util.concurrent` package of the `java.base` module. The `Executors` class defines static methods such as `newFixedThreadPool`.

Sample code

Let's imagine a simple straightforward example that is easily applicable in the real world using a suitable abstraction. The goal is to design an application that tracks vehicle production. Most likely, the company offers different types of vehicles. Each vehicle can be represented by its own object. To draw the intent, we created a **Unified Modeling Language** (**UML**) class diagram to maintain clarity (*Figure 3.2*):

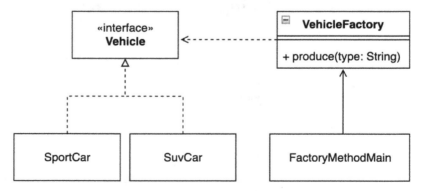

Figure 3.2 – The vehicle production tracking example

The intended factory intends to produce two different types of vehicles, and the application fulfills the wishes on the fly (*Example 3.1*):

```java
public static void main(String[] args) {
    System.out.println("Pattern Factory Method: Vehicle
        Factory 2");
    var sportCar = VehicleFactory.produce("sport");
    System.out.println("sport-car:" + sportCar);
    sportCar.move();
}
```

Here is the output:

```
Pattern Factory Method: Vehicle Factory 2
sport-car:SportCar[type=porsche 911]
SportCar, type:'porsche 911', move
```

Example 3.1 – VehicleFactory produces vehicles of the same "family" based on input arguments

Instead of distributing the creation of such vehicle types over multiple places, we create a factory. The factory abstraction concentrates the whole vehicle composition process and exposes only one entry point that allows the client to create the desired vehicle type (as shown in *Example 3.2*). A factory only implements one static method, so it makes sense to keep its constructor private because factory instances are undesirable:

```java
final class VehicleFactory {
private VehicleFactory(){}
    static Vehicle produce(String type){
        return switch (type) {
            case "sport" -> new SportCar("porsche 911");
            case "suv" -> new SuvCar("skoda kodiaq");
            default -> throw new
                IllegalArgumentException("""
            not implemented type:'%s'
                """.formatted(type));
        };
```

```
        }
    }
```

Example 3.2 – The VehicleFactory class exposes the static factory method to produce an instance of the object that implements a Vehicle interface

The presented `switch` expression may use the pattern-matching approach to simplify code instead of the traditional `switch-label-match` construct. The application provides multiple implementations of the vehicle (*Example 3.3*):

```
interface Vehicle {
    void move();
}
```

Example 3.3 – Each considered vehicle inherits the method abstractions through the Vehicle interface

Due to another platform-syntactical improvement, the `records` type, it is possible to choose the level of the class encapsulation with the reflection of SOLID principles. It depends on how much the software architect intends to allow the vehicle instance to change its inner state. Let us first look at the standard Java class definition approach (*Example 3.4*):

```
class SuvCar implements Vehicle {
    private final String type;
    public SuvCar(String t){
        this.type = t;
    }
    @Override
    public void move() {...}
}
```

Example 3.4 – SuvCar allows adding inner fields that could hold a mutable state

The software architect has the chance to use the `record` class to create immutable instances of the desired vehicle along with the `hashCode` and `equals` methods followed by the `toString` implementation:

```
record SportCar(String type) implements Vehicle {
    @Override
    public void move() {
        System.out.println("""
```

```
            SportCar, type:'%s', move""".formatted(type));
    }
}
```

Example 3.5 – SportCar is considered to be immutable

The recently introduced `record` feature reduces the potential boilerplate code base while still allowing internal functionality to be implemented (as discussed in the *Records (Java SE 16, JEP-395)* section of the previous chapter).

Conclusion

A factory method has some limitations. The most important one is that it can only be used for a specific family of objects. This means that all classes must maintain similar properties or common ground. The deviation from the base class of a class can introduce a dramatic strong coupling between the code and the application.

The point to consider may be related to the method itself, as it may be static or belong to the instances (as covered in the previous chapter under *The stack area* and *The heap area* sections, respectively). This depends on the software designer's decision.

The object of the one family is created. Let us investigate how to deal with factory families that share a common property.

Creating objects from different families using the abstract factory pattern

This pattern introduces a factory abstraction without the requirement to define specific classes (or classes that should be instantiated). The client requests a proper factory that instantiates the object instead of attempting to create it. The abstract factory pattern was mentioned in the GoF's book.

Motivation

Modularizing applications can become a challenge. Software designers can avoid adding code to classes to preserve encapsulation. The motivation is to separate the factory logic from the application code so that it can supply the appropriate factory to produce the required objects. An abstract factory provides a standardized way to create an instance of the desired factory and deliver that instance to the client for use. The client uses the resulting factory to instantiate the object. Abstract factory provides an interface for creating both factories and objects without specifying their classes. The pattern implicitly supports SOLID principles and maintainability by isolating the logic of participants and insiders. The application is independent of how its products are created, composed, and represented.

Finding it in the JDK

The abstract factory method pattern can be found in the JDK in the java.xml package of the java. xml module. The abstract factory pattern can be found in the representation and implementation of the DocumentBuilderFactory abstract class and its static newInstance method. The factory uses a lookup service to find the required builder implementation.

Sample code

Consider that although vehicles share some common features, their production requires different kinds of processes (*Figure 3.3*):

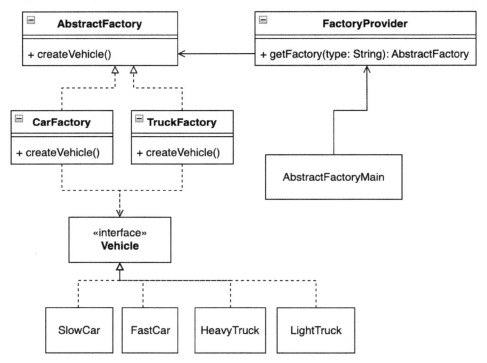

Figure 3.3 – Manufacturing different types of vehicles with the Abstract Factory pattern

In such cases, we create multiple factories responsible for specific objects. Although these classes belong to different families, they do have properties in common. An important feature is that each factory can implement its own initialization sequence while sharing generic logic. The example requires the correct CarFactory instance to create a SlowCar object (*Example 3.6*):

```
public static void main(String[] args) {
    ...
```

```
    AbstractFactory carFactory =
        FactoryProvider.getFactory("car");
    Vehicle slowCar = carFactory.createVehicle("slow");
        slowCar.move();
}
```

Here is the output:

```
Pattern Abstract Factory: create factory to produce
    vehicle...
slow car, move
```

Example 3.6 – The client decides which Vehicle type is required

A key element of the game is the factory provider; this distinguishes which factory is created based on incoming arguments (*Example 3.7*). The provider is implemented as a utility, so its class is final, and the constructor is private because instances are not required. Of course, the implementation may vary depending on the requirements:

```
final class FactoryProvider {
private FactoryProvider(){}
    static AbstractFactory getFactory(String type){
        return switch (type) {
            case "car" -> new CarFactory();
            case "truck" -> new TruckFactory();
            default -> throw new IllegalArgumentException
                ("""          this is %s
                """.formatted(type));
        };
    }
}
```

Example 3.7 – The FactoryProvider class defines how the particular factory of the object families is configured and instantiated

Each factory from the group shares common logic or features to maintain the DRY approach in the code base:

```
abstract class AbstractFactory {
    abstract Vehicle createVehicle(String type);
}
```

Example 3.8 – The AbstractFactory class provides the common logic or methods that may require implementation by a specific factory

These individual factories can implement additional logic to distinguish which product should be delivered, similar to the `TruckFactory` and `CarFactory` implementations in the following example (*Example 3.9*):

```
class TruckFactory extends AbstractFactory {
    @Override
    Vehicle createVehicle(String type) {
        return switch(type) {
            case "heavy" -> new HeavyTruck();
            case "light" -> new LightTruck();
            default -> throw new IllegalArgumentException
                ("not implemented");
        };
    }
}
```

Example 3.9 – The TruckFactory class represents the specific AbstractFactory implementation

Conclusion

The abstract factory pattern provides consistency across products. Using a super factory can cause instability in the client runtime because the requested product can throw an exception or error due to incorrect implementation, as such information was not known on the fly. The abstract factory pattern, on the other hand, promotes testability. An abstract factory is free to represent the many other interfaces that came with its implementation. The pattern provides a common way to deal with products without depending on their implementation, which can improve the separation of concerns of application code. It can use interfaces or abstract classes. The client becomes independent of how objects are composed and created.

The benefit of encapsulating factories and code separation can be seen as a limitation. An abstract factory must be controlled by one or more parameters to properly define a dependency. To improve the code maintainability of the required factories, it may be useful to consider the previously discussed *sealed classes* enhancements (see the *Sealed classes (Java SE 17, JEP-409)* section in the previous chapter). Sealed classes can have a positive impact on code base stability.

Let us examine how to customize an object creation process in the next section.

Instantiating complex objects with the builder pattern

The builder pattern helps separate the construction of a complex object from its code representation so that the same composition process can be reused to create different configurations of an object type. The builder design pattern was identified early and is the part of GoF's book.

Motivation

The main motivation behind the builder pattern is to construct complex instances without polluting the constructor. It helps to separate or even break down the creation process into specific steps. The composition of objects is transparent to the client and allows the creation of different configurations of the same type. The builder is represented by a separate class. It can help to transparently extend the constructor on demand. The pattern helps to encapsulate and enforce the clarity of the instantiation process with respect to the previously discussed SOLID design principles.

Finding it in the JDK

The builder pattern is a commonly used pattern inside the JDK. A great example is creating sequences of characters that represent a string. For example, `StringBuilder` and `StringBuffer` are located in the `java.lang` package of the `java.base` module, which is visible by default to every Java application. The string builder provides multiple overloaded connection methods that accept different types of input. Such input is concatenated with an already-created byte array. Another example can be found in the `java.net.http` package represented by the `HttpRequest.Builder` interface and its implementation or the `Stream.Builder` interface found in the `java.util.stream` package. As mentioned earlier, the builder pattern is very often used. Worth noting are `Locale.Builder` and `Calendar.Builder`, which use the setter methods to store values of the final product. Both can be found in the `java.util` package of the `java.based` module.

Sample code

The builder, which is the key element of the pattern, holds required field values during a `Vehicle` instance creation, more precisely, references to the objects (*Figure 3.4*):

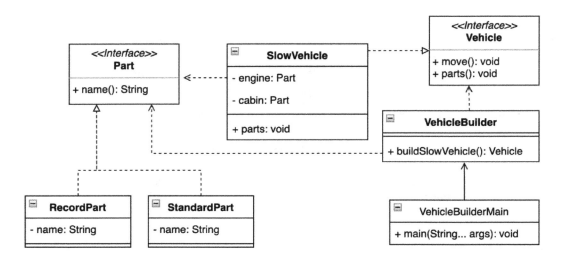

Figure 3.4 – How to transparently make a new vehicle with the builder pattern

The overall responsibility of the builder pattern is to create vehicles (*Example 3.10*):

```
public static void main(String[] args) {
    System.out.println("Builder pattern: building
        vehicles");

    var slowVehicle = VehicleBuilder.buildSlowVehicle();
    var fastVehicle = new FastVehicle.Builder()
                        .addCabin("cabin")
                        .addEngine("Engine")
                        .build();
    slowVehicle.parts();
    fastVehicle.parts();
}
```

Here is the output:

```
Builder pattern: building vehicles
SlowVehicle,engine: RecordPart[name=engine]
SlowVehicle,cabin: StandardPart{name='cabin'}
FastVehicle,engine: StandardPart{name='Engine'}
FastVehicle,cabin: RecordPart[name=cabin]
```

Example 3.10 – The builder pattern can be implemented in several ways, depending on the requirements

The builder pattern may be implemented using different approaches. One approach is to encapsulate and hide all builder logic and provide a product directly without exposing implementation details:

```
final class VehicleBuilder {
    static Vehicle buildSlowCar(){
        var engine = new RecordPart("engine");
        var cabin = new StandardPart("cabin");
        return new SlowCar(engine, cabin);
    }
}
```

Example 3.11 – The VehicleBuilder hides the logic in order to provide a particular instance

Or the builder may become a part of the class for which the instance is intended to be created. In such a case, it is possible to decide which element should be added to the newly created specific instance:

```
class FastCar implements Vehicle {
    final static class Builder {
        private Part engine;
        private Part cabin;
        Builder(){}
        Builder addEngine(String e){...}
        Builder addCabin(String c){...}
        FastCar build(){
            return new FastCar(engine, cabin);
        }
    }

    private final Part engine;
    private final Part cabin;
    ...
    @Override
    public void move() {...}
    @Override
```

```
        public void parts() {...}
  }
```

Example 3.12 – The FastVehicle.Builder is represented as a static class and needs to be instantiated, and provides the possibility of final result customization

Both example approaches are implemented according to SOLID principles. The builder pattern is a nice example of abstraction, polymorphism, inheritance, and encapsulation (APIE) principles and is very open to refactoring, extending, or validating properties.

Conclusion

The builder pattern helps enforce the single responsibility principle by separating complex creation from business logic. It also improves code readability and the DRY principle, as the instantiation is extensible and user-understandable. The builder pattern is a commonly used design pattern because it reduces "code smell" and constructor pollution. It also improves testability. The code base helps avoid multiple constructors with different representations, some of which have never been used.

Another good point to consider while implementing a pattern is to use the JVM's heap or stack – more specifically, to create a statically or dynamically allocated representation of the pattern. This decision is commonly answered by the software designers themselves.

It is not always necessary to reveal the construction process. The next section presents the simplicity of object cloning.

Cloning objects with the prototype pattern

The prototype pattern solves the difficulty of creating new instances of an object with complicated instantiation process, which is too cumbersome and undesirable because it can lead to unnecessary subclassing. The prototype is a very common design pattern and was described in the GoF's book.

Motivation

The prototype design pattern becomes very useful when heavy objects need to be created and factories are an unwanted approach. The newly created instance is cloned from its parent because the parent acts as a prototype. Instances are independent of each other and can be customized. Instance logic is not exposed to, and cannot be contributed to by, the client.

Finding it in the JDK

There are many examples of using prototype patterns across JDK packages. The Collection framework members implement the `clone` method required by the inherited `Cloneable` interface. For example, an `ArrayList.clone()` method execution creates a shallow `List` copy of entities, field by field.

Another prototype implementation could be the `Calendar` class from the `java.util` package of the `java.base` module. A clone of the overridden method is also used for the `Calendar` implementation itself, as it helps to avoid unwanted modification of an already configured one. Usage can be found in the `getActualMinimum` and `getActualMaximum` methods.

Sample code

When there are only a few vehicle models in production, there is no need to constantly establish new objects by factories or builders, which could actually lead to unwieldy code behavior as internal properties may change. Imagine the early stage of vehicle production where equality is required with each new iteration to track progress (*Figure 3.5*):

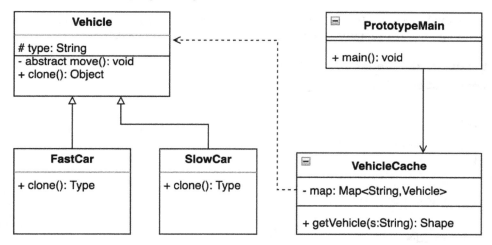

Figure 3.5 – Creating new instances from the prototype

In such cases, it is easier to create an exact copy of an already designed vehicle as its prototype:

```java
public static void main(String[] args) {
    Vehicle fastCar1 = VehicleCache.getVehicle("fast-car");
    Vehicle fastCar2 = VehicleCache.getVehicle("fast-car");
    fastCar1.move();
    fastCar2.move();
    System.out.println("equals : " + (fastCar1
        .equals(fastCar2)));
}
```

Here is the output:

```
Pattern Prototype: vehicle prototype 1
fast car, move
fast car, move
equals : false
fastCar1:FastCar@659e0bfd
fastCar2:FastCar@2a139a55
```

Example 3.13 – A new vehicle can be cloned from the available instances

The instances can be recreated (respectively, cloned) on demand. The Vehicle abstract class provides a foundation for each new prototype implementation and provides cloning details:

```
abstract class Vehicle implements Cloneable{
    protected final String type;

    Vehicle(String t){
        this.type = t;
    }

    abstract void move();

    @Override
    protected Object clone() {
        Object clone = null;
        try{
        clone = super.clone();
        } catch (CloneNotSupportedException e){...}
        return clone;
    }
}
```

Example 3.14 – The Vehicle abstract class must implement a Cloneable interface and introduce the clone method implementation

Each vehicle implementation requires the extension of the parent `Vehicle` class:

```
class SlowCar extends Vehicle {
    SlowCar(){
        super("slow car");
    }
    @Override
    void move() {...}
}
```

Example 3.15 – The Vehicle interface-specific implementation provided by the SlowCar class and the move method implementation

The prototype pattern introduces an internal cache that collects available `Vehicle` type prototypes (*Example 3.16*). The proposed implementation implements a static method to make the cache work as a tool. It makes sense for its constructor to be private:

```
final class VehicleCache {
private static final Map<String, Vehicle> map =
    Map.of("fast-car", new FastCar(), "slow-car", new
        SlowCar());

private VehicleCache(){}
    static Vehicle getVehicle(String type){
        Vehicle vehicle = map.get(type);
        if(vehicle == null) throw
        new     IllegalArgumentException("not allowed:" +
            type);
        return (Vehicle) vehicle.clone();
    }
}
```

Example 3.16 – VehicleCache holds the references to the already prepared prototypes that may be cloned

The examples show that the client works each time with an identical copy of the base prototype. This copy may be customized based on the requirements.

Conclusion

The prototype pattern is useful for dynamic loading or to avoid increasing the complexity of the code base by introducing unnecessary abstractions, known as **subclassing**. This does not mean that clones do not need to implement the interface, but cloning can reduce exposure requirements or make the instantiation process too complicated. A prototype properly encapsulates the complicated logic of an instance that is not meant to be touched or modified. Software designers should be aware of the possibility that such a code base can easily change to legacy code. On the other hand, a pattern can defer and support iterative changes to the code base.

Multiple instances of an object are not always desirable and are sometimes even undesirable. In the next section, we'll learn how to guarantee the presence of only one unique class instance at runtime.

Ensuring only one instance with the singleton pattern

A singleton object provides transparent and global access to its instance and ensures that only one instance is present. The singleton pattern was identified very early by industry requirements and is mentioned in the GoF's book.

Motivation

A client or application wants to ensure that only one instance is present at runtime. An application may require multiple object instances that all use one unique resource. This fact introduces instability because any of these objects can access such a resource. A singleton guarantees only one instance that provides a global access point to all clients within the desired scope of the running JVM.

Finding it in the JDK

The best example of using a singleton is a running Java application, or more precisely, the runtime. It is found in the Runtime class and its method, getRuntime, resides in the java.lang package of the java.base module. The method returns an object associated with the current Java application. The runtime instance allows the client to add, for example, shutdown hooks to the running application.

Sample code

The following example suggests an application that only runs one car with its engine (*Figure 3.6*):

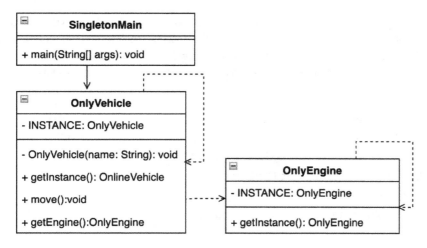

Figure 3.6 – How the Singleton pattern represents an engine

In other words, this means that only one instance of a particular type of engine and vehicle must be present in the JVM:

```
public static void main(String[] args) {
    System.out.println("Singleton pattern: only one
        engine");
    var engine = OnlyEngine.getInstance();
    var vehicle = OnlyVehicle.getInstance();
    vehicle.move();
    System.out.println("""
        OnlyEngine:'%s', equals with vehicle:'%s'"""
        .formatted(engine, (vehicle.getEngine()
            .equals(engine))));
}
```

Here is the output:

```
Pattern Singleton: only one engine
OnlyVehicle, move
OnlyEngine:'OnlyEngine@7e9e5f8a', equals with
    vehicle:'true'
```

Example 3.17 – One instance of OnlyEngine and OnlyCar is present at runtime

There are a couple of different ways to ensure that the instance of the object will be unique. The implementation of the `OnlyEngine` class introduces the possible singleton implementation where its instance is lazily created on demand (*Example 3.18*). The `OnlyEngine` class implements the generic `Engine` interface. Its implementation provides a static `getInstance` method as a transparent entry point:

```
interface Engine {}
class OnlyEngine implements Engine {
    private static OnlyEngine INSTANCE;
    static OnlyEngine getInstance(){
        if(INSTANCE == null){
            INSTANCE = new OnlyEngine();
        }
        return INSTANCE;
    }
    private OnlyEngine(){}
}
```

Example 3.18 – The OnlyEngine class checks the existence of its instance – lazy initiation

Another way to implement a singleton is to create a `static` field that belongs to the class itself and expose a `getInstance` entry point to the potential client (*Example 3.19*). It is fair to note that, in such a case, the constructor becomes private:

```
class OnlyVehicle {
    private static OnlyVehicle INSTANCE = new
        OnlyVehicle();
    static OnlyVehicle getInstance(){
        return INSTANCE;
}
    private OnlyVehicle(){
        this.engine = OnlyEngine.getInstance();
    }
    private final Engine engine;
    void move(){
        System.out.println("OnlyVehicle, move");
    }
    Engine getEngine(){
```

```
            return engine;
    }
}
```

Example 3.19 – The OnlyVehicle class provides its instance as a static field that belongs to the class

The lazily initialized singleton pattern implementation may become a challenge in a multithreaded environment where the `getInstance` method must be synchronized to obtain a unique instance. One possibility is to create a singleton as an `enum` class (*Example 3.20*):

```
enum OnlyEngineEnum implements Engine {
    INSTANCE;
    }
    ...
    private OnlyVehicle(){
    this.engine = OnlyEngineEnum.INSTANCE;
}
...
```

Example 3.20 – The OnlyEngineEnum singleton enum class approach

Conclusion

Singleton is a relatively trivial design pattern, although it can get complicated when used inside a multithreading environment due to its guarantee that there will be only one instance of the required object. The pattern can be challenged when enforcing the principle of single responsibility since the class is actually responsible for instantiating itself. On the other hand, the singleton pattern ensures that clients can access allocated resources globally, preventing accidental object initialization or destruction. The pattern should be used wisely as it creates tightly coupled code in the manner of required class instantiations, which can tend to cause testability issues. The pattern also suppresses other subclasses, making any extension nearly impossible.

Creating instances is not always a good approach. Let us examine how to do this on demand.

Improving performance with the object pool pattern

The object pool pattern instantiates ready-to-use objects and limits their initialization time. The required instances can be recreated on demand. A pool of objects can represent a base of conditions on which new instances could be created, or limit their creation.

Motivation

Instead of constantly creating new object instances over the code base, the object pool provides an encapsulated solution for managing application or client performance by serving an already initialized object ready for use. The pattern separates the build logic from the business code and helps manage the resource and performance aspects of the application. It may not only help with an object's life cycle but also with validation when it is created or destroyed.

Finding it in the JDK

A nice example of the object pool pattern is the `ExecutorService` interface found in the `java.util.concurrent` package and the implementation provided by the `Executors` class of the `util` factory that handles the appropriate executor instances, for example, the `newScheduledThreadPool` method.

Sample code

The current example introduces a scenario where the garage contains a specific number of cars that drivers may drive (*Figure 3.7*):

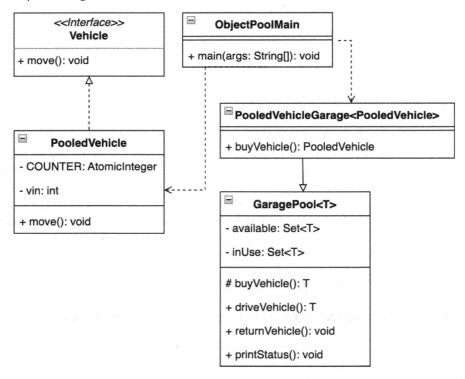

Figure 3.7 – A garage following the Object Pool pattern

When a car is not available, the garage has implemented a logic to buy a new one in order to keep all drivers busy:

```
public static void main(String[] args) {
    var garage = new PooledVehicleGarage();
    var vehicle1 = garage.driveVehicle();
    ...
    vehicle1.move();
    vehicle2.move();
    vehicle3.move();
    garage.returnVehicle(vehicle1);
    garage.returnVehicle(vehicle3);
    garage.printStatus();
    var vehicle4 = garage.driveVehicle();
    var vehicle5 = garage.driveVehicle();
    vehicle4.move();
    vehicle5.move();
    garage.printStatus();
}
```

Here is the output:

```
Pattern Object Pool: vehicle garage
PooledVehicle, move, vin=1
PooledVehicle, move, vin=2
PooledVehicle, move, vin=3
returned vehicle, vin:1
returned vehicle, vin:3
Garage Pool vehicles available=2[[3, 1]] inUse=1[[2]]
PooledVehicle, move, vin=3
PooledVehicle, move, vin=1
Garage Pool vehicles available=0[[]] inUse=3[[3, 2, 1]]
```

Example 3.21 – The pooling vehicles in the garage instance help reduce the cost of the object

A core element of the pattern is pool abstraction, as it contains all the required logic for managing entities. One option is to create an abstract garage pool class (*Example 3.22*) that contains all the synchronization mechanisms. These mechanisms are required to avoid potential code instability and inconsistency:

```java
abstract class AbstractGaragePool<T extends Vehicle> {
    private final Set<T> available = new HashSet<>();
    private final Set<T> inUse = new HashSet<>();
    protected abstract T buyVehicle();
    synchronized T driveVehicle() {
        if (available.isEmpty()) {
            available.add(buyVehicle());
        }
        var instance = available.iterator().next();
        available.remove(instance);
        inUse.add(instance);
        return instance;
    }
    synchronized void returnVehicle(T instance) {...}
    void printStatus() {...}
}
```

Example 3.22 – An abstract garage pool provides all the required logic to properly administrate elements

The garage pool limits the possible instance types. A class is bound by the Vehicle abstraction (*Example 3.23*). The interface provides the common functions used by the client. In the following example, the implementation of AbstractGaragePool represents the client:

```java
interface Vehicle {
    int getVin();
    void move();
}
```

Example 3.23 – The Vehicle interfaces need to be implemented by PooledVehicle

In addition to implementing functions, the `PooledVehicle` class provides a private counter (*Example 3.24*). A counter belongs to a class, so it is marked as `static` and `final`. The counter counts the number of instances purchased by the garage pool:

```
class PooledVehicle implements Vehicle{
    private static final AtomicInteger COUNTER = new
        AtomicInteger();

    private final int vin;
    PooledVehicle() {
        this.vin = COUNTER.incrementAndGet();
    }

    @Override
    public int getVin(){...}

    @Override
    public void move(){..}
}
```

Example 3.24 – The PooledVehicle class implementation also holds the number of instances created

Conclusion

Improving client performance helps reduce the expensive object instantiation time (as we saw in *Example 3.21*). Object pools are also quite useful in cases where only short-lived objects are required, as they help reduce memory fragmentation by uncontrolled instances. It is worth touching upon the implementation of the internal cache pattern, as seen in the garage example.

Although the pattern is quite efficient, the right choice of collection structure can also have a dramatic impact on its performance. It can reduce search and save time.

Another positive outcome can be considered as the impact on the garbage collection process and memory compaction due to the analysis of active objects since there may be fewer objects to analyze.

It is not always necessary to store everything in memory for later reuse. Let us examine how to postpone the object initialization process and not polluting memory.

Initiating objects on demand with the lazy initialization pattern

This pattern's purpose is to defer an instance of the desired class instance until the client actually requests it.

Motivation

Although operational memory has grown drastically over the years, we learned in the previous chapter that the JVM allocated a defined, specific size of memory reserved for the heap. When the heap is exhausted and the JVM is unable to allocate any new object, it causes an out of memory error. Lazy handling can have quite a positive impact on this heap pollution. It is sometimes also called asynchronous loading because of the delayed instance. The pattern has quite a nice use in a web application where the web page is generated on demand rather than during the application initialization process. It also has its place in an application, where the cost of operating the relevant object is high.

Finding it in the JDK

Lazy initialization can be demonstrated using the example of dynamic loading by `ClassLoader` of classes that were not linked at runtime when the application was started. Classes can be loaded eagerly or deferred by the class policy. Certain classes, such as `ClassNotFoundException`, are loaded implicitly through the `java.base` module. They support the class implementation located in the `java.lang` package and its `forName` method. The implementation of the method is provided by an internal API. A lazy initiated class may be the reason the application requires a warm-up time. For example, `Enum` classes are a special type of static final classes that act as constants and are loaded eagerly.

- > refers to loading steps into the class loader and filling out the appropriate method area, as we learned in the previous chapter.

Sample code

The basic idea of the lazy initialization example is that the created vehicle is initialized on demand or, when already created, a reference is provided to the client (*Figure 3.8*):

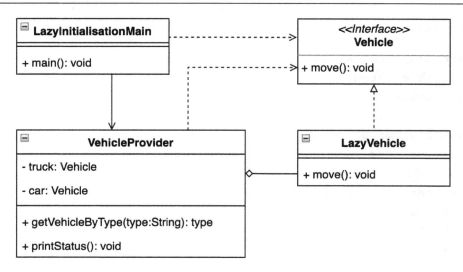

Figure 3.8 – How to create a vehicle on demand with the lazy initiation pattern

Such vehicles are present only if they are really required by the client. In such cases, specific vehicle instances are created. When a vehicle is already present inside the provider context, then such an instance is reused:

```
public static void main(String[] args) {
    System.out.println("Pattern Lazy Initialization: lazy
        vehicles");
    var vehicleProvider = new VehicleProvider();
    var truck1 = vehicleProvider.getVehicleByType("truck");
    vehicleProvider.printStatus();
    truck1.move();
    var car1 = vehicleProvider.getVehicleByType("car");
    var car2 = vehicleProvider.getVehicleByType("car");
    vehicleProvider.printStatus();
    car1.move();
    car2.move();
    System.out.println("ca1==car2: " + (car1.equals
        (car2)));
}
```

Here is the output:

```
Pattern Lazy Initialization: lazy vehicles
lazy truck created
status, truck:LazyVehicle[type=truck]
status, car:null
LazyVehicle, move, type:truck
lazy car created
status, truck:LazyVehicle[type=truck]
status, car:LazyVehicle[type=car]
LazyVehicle, move, type:car
LazyVehicle, move, type:car
ca1==car2: true
```

Example 3.25 – An example implementation of the pooled vehicle

The implementation of the `VehicleProvider` class considers the fields private. Those fields hold the references to the desired vehicle if required. The provider encapsulates the decision and instantiation logic. One of the possible implementations may use the `switch-label-match` construct (*Example 3.26*), `switch` expressions, and so on. It's good to point out that in this example, the `VehicleProvider` class requires an instance in the package scope, so its constructor is a `private package` and is not exposed to other packages:

```
final class VehicleProvider {
    private Vehicle truck;
    private Vehicle car;
    VehicleProvider() {}
    Vehicle getVehicleByType(String type){
        switch(type){
        case "car":
            ...
            return car;
        case "truck":
            if(truck == null){
                System.out.println("lazy truck created");
                truck = new LazyVehicle(type);
            }
            return truck;
```

```
        default:
            ...
    }
    void printStatus(){...}
}
```

Example 3.26 – VehicleProvider hides the possible entity instantiation logic from the client

In order to enforce the extendibility to the possible lazily initiated object, each entity implements the `Vehicle` abstraction:

```
interface Vehicle {
    void move();
}

record LazyVehicle(String type) implements Vehicle{
    @Override
    public void move() {
        System.out.println("LazyVehicle, move, type:" +
            type);
    }
}
```

Example 3.27 – Vehicle abstraction and possible LazyVehicle implementation using record to enforce immutability

Such an approach enforces continual vehicle evolution without complicated changes to the provider logic.

Conclusion

The lazy initialization design pattern can help keep application memory small. Improper use, on the other hand, can cause unwanted delays, as objects can be too complex to create and take a significant amount of time to run.

The next section shows how to inject the logic into the client, represented by the newly created vehicle instance.

Reducing class dependencies with the dependency injection pattern

This pattern separates the initialization of the class (that *acts* as a service) from the client (that *uses* the service).

Motivation

The dependency injection pattern is widely used where there is a need to separate the implementation of a particular object (service) from the target object (client) that uses its exposed services, methods, and the like. Services are available when a client instance is to be created. The pattern allows you to eliminate any hardcoded dependencies. These services are instantiated outside of the client creation process. This means that the two are loosely connected, and SOLID principles can be enforced. There are three ways that dependency injection can be implemented:

- **Constructor dependency injection**: Intended services are made available to the client through the initialization of the constructor.

- **Injection method**: The client exposes the method normally through an interface. Such a method supplies dependencies to the client. The supplier object uses a method to inject the service(s) into the client.

- **Field dependency injection**: This type of injection is done using a setup-like method. These setters refer to the respective field held by the client. The client can also expose the field as a `public` property.

Finding it in the JDK

A good example of using the dependency injection pattern is the `ServiceLoader` utility class. It can be found in the `java.base` module and its `java.util` package. The `ServiceLoader` instance tries to find services during the application startup at runtime. A service is considered to be represented by a well-specified interface that is implemented by the relevant service provider or providers. The application code is able to distinguish the desired provider at runtime. It is good to note that `ServiceLoader` works with the classic `classpath` configuration or can be used seamlessly with the Java Platform Module System (discussed in *Chapter 2, Discovering the Java Platform for Design Patterns*).

In the past, dependency injection was part of the Java EE scope, not the classic JDK. This means that the feature was available on the Java platform. Due to the evolution of the JDK, the dependency injection features were moved to the **Jakarta Dependency Injection** project. A newly created project follows its own release and development cycle without dependency on the JDK. Nevertheless, the dependency injection is well known due to the usage of @Inject, @Named, @Scope, or @Qualifier annotations. These annotations allow the class to be turned into a managed object at runtime, where the implementation of the desired provider can be distinguished.

Sample code

An example shows a simplified dependency injection pattern in order to receive a general experience. It is a very trivial implementation; in fact, it draws a picture of how the previously mentioned API works behind the scenes. Let's imagine a scenario where the intended vehicle instance requires an engine (*Figure 3.9*):

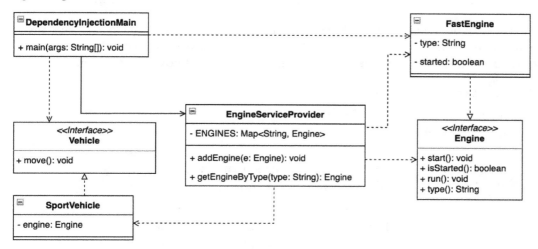

Figure 3.9 – Injecting engine as a service to the vehicle instance

Specific engine construction logic is detached from vehicle-related code (*Example 3.28*):

```
1 public static void main(String[] args) {
2     System.out.println("Pattern Dependency Injection:
          vehicle and engine");
3     EngineServiceProvider.addEngine(new FastEngine
          ("sport"));
4     Engine engine =
          EngineServiceProvider.getEngineByType("sport");
5     Vehicle vehicle = new SportVehicle(engine);
6     vehicle.move();
7 }
```

Here is the output:

```
Pattern Dependency Injection: vehicle and engine
FastEngine, started
FastEngine, run
SportCar, move
```

Example 3.28 – Creating an instance of FastEngine separately from the vehicle and then adding the engine to the constructed SportVehicle

The `FastEngine` instance is used when it is fully ready (initiated, verified, and so on). The desired type of vehicle can be constructed independently without any dependence on engine logic. An engine instance is provided to `SportVehicle` using `EngineServiceProvider` (*Example 3.29*):

```java
final class EngineServiceProvider {
    private static final Map<String, Engine> ENGINES = new
        HashMap<>();
    ...
    static Engine getEngineByType(String t){
        return ENGINES.values().stream()
                .filter(e -> e.type().equals(t))
                .findFirst().orElseThrow
                    (IllegalArgumentException::new);
    }
}
```

Example 3.29 – EngineServiceProvider registers instantiated reusable services

The `SportVehicle` class implements the `Vehicle` interface (*Example 3.30*) in order to reflect the open-close approach mentioned as part of the SOLID design principles:

```java
interface Vehicle {
    void move();
}
class SportVehicle implements Vehicle{
    private final Engine engine;
    SportVehicle(Engine e) {...}
    @Override
    public void move() {
```

```
            if(!engine.isStarted()){
                engine.start();
        }

            engine.run();
            System.out.println("SportCar, move");
        }
}
```

Example 3.30 – SportVehicle implements the Vehicle interface together with additional internal logic for the provided Engine instance

It is important to note that although a specific Engine type (*Example 3.31*) instance (FastEngine) is created somewhere else (*Example 3.28*, line 3), its presence is required when the SportVehicle object is instantiated (*Example 3.28*, line 5):

```
interface Engine {
    void start();
    boolean isStarted();
    void run();
    String type();
}
class FastEngine implements Engine{
    private final String type;
    private boolean started;
    FastEngine(String type) {
        this.type = type;
    }
    ...
}
```

Example 3.31 – Engine interfaces implemented by the FastEngine class and provided by EngineServiceProvider

The key player object in the described example is EngineServiceProvider. It provides a reference to already created desired Engine instances and distributes them across the business code. This means that any client that needs to work on Engine, similar to the SportVehicle instance, will get access to the correct instance through the link exposed by EngineServiceProvider.

The presented trivial example can be easily turned into another using the `ServiceProvider` utility class instance. The changes are very minimal (*Example 3.32*):

```
public static void main(String[] args) {
    System.out.println("Pattern Dependency Injection
        Service Loader: vehicle and engine");
    ServiceLoader<Engine> engineService =
        ServiceLoader.load(Engine.class);
    Engine engine = engineService.findFirst()
        .orElseThrow();
    Vehicle vehicle = new SportVehicle(engine);
    vehicle.move();
}
```

Example 3.32 – ServiceLoader provides an available implementation of the Engine interface used to instantiate the SportVehicle type

In the standard classpath utilization, the Java platform requires service providers to be registered in the `META-INF` folder and the `services` subfolder. The filename is made up of the package and the service interface name, and the file contains the available service providers.

The Java Platform Module System simplifies the configuration steps. The relevant modules provide (and make available service implementations to) the target modules, as we touched on in *Chapter 2, Discovering the Java Platform for Design Patterns*.

Conclusion

The dependency injection pattern ensures that the client does not know about the instantiation of the used service. The client has access to the service through the common interfaces. It makes the code base better testable. It also simplifies the code base testability. Dependency injection is a widely used pattern by various frameworks, such as Spring and Quarkus. Quarkus uses the Jakarta Dependency Injection specification. The dependency injection pattern conforms to the SOLID and APIE object-oriented programming principles as it provides the abstraction of interfaces. The code does not depend on the implementation but communicates with the services through the interfaces. The dependency injection pattern enforces a DRY principle as it is not required to continually initiate a service.

Summary

Creational design patterns play a very important role in software application design. They help in transparently centralizing object instantiation logic with respect to basic object-oriented principles. The examples showed that each pattern may have multiple implementations. This is because the implementation decision may depend on other software architecture factors. Those factors take into account JVM heap and stack usage, application runtime, or business logic encapsulation.

Using design patterns implicitly for authoring promotes a DRY approach, which has a positive impact on application development and reduces code base pollution. The application becomes testable and software architects have a framework to confirm the presence of expected objects inside the JVM. This becomes particularly important when a logic issue is identified, which could be an exception or unexpected results. A well-created code base helps to get into the root cause quite fast, maybe even without the necessity to debug.

In this chapter, we have learned how to create an object of a particular family using the transparent factory method pattern. The abstract factory pattern showed the encapsulated way of creating different factories types. Not all the required information is always present at one moment, and the builder pattern introduced a way to deal with such a challenge of composing complex objects. The prototype pattern showed the approach of not exposing the instance logic to the client, where the singleton pattern ensured the presence of a single instance at runtime. The object pool design pattern revealed how to improve memory usage at runtime, while the lazy instantiation pattern showed how to defer an object until it is needed. The dependency injection pattern demonstrated the reusability of instances when creating new objects.

Creational design patterns not only bring clarity to the creation of new instances but, in many cases, they can also contribute to the decision about the correct structure of the code (a code structure that reflects the required business logic). The last three mentioned patterns solve not only the issue of creating objects but also their reusability for efficient memory use. The examples we covered showed us how creational patterns can be implemented and we learned about the differences and purposes of each of them.

The next chapter will discuss the structure design pattern. This pattern will help us organize our code based on the most common scenarios. Let's move on.

Questions

1. What challenges do the creational design patterns solve?
2. Which patterns may be helpful to reduce object initiation costs?
3. What is the key reason to utilize the singleton design pattern?
4. Which pattern helps to reduce constructor pollution?
5. How do you hide complex instantiation logic from the client?

6.   Is it possible to reduce the instantiation application memory footprint?

7.   What design pattern is useful for creating objects of a specific family?

Further reading

- *Design Patterns: Elements of Reusable Object-Oriented Software* by Erich Gamma, Richard Helm, Ralph Johnson, and John Vlissides, Addison-Wesley, 1995

- *Design Principles and Design Patterns* by Robert C. Martin, Object Mentor, 2000

- *Cramming more components onto integrated circuits* by Gordon E. Moore, Electronics Magazine, 1965-04-19

- *Oracle Tutorials: Generics*: `https://docs.oracle.com/javase/tutorial/java/generics/index.html`

- *Quarkus Framework*: `https://quarkus.io/`

- *Spring Framework*: `https://spring.io/`

- *Jakarta Dependency Injection*: `https://jakarta.ee/specifications/dependency-injection/`

- *Clean Code* by Robert C. Martin, Pearson Education, Inc, 2009

- *Effective Java – Third Edition* by Joshua Bloch, Addison-Wesley, 2018

4

Applying Structural Design Patterns

Every piece of software has a purpose or, in other words, an expected behavior that it should fulfill. While the previous chapter described in detail creational design patterns, this chapter will focus on designing maintainable and flexible source code for objects created. Structural patterns attempt to bring clarity to relationships between created instances, not only to maintain an application but also to easily understand its purpose. Let us dive deeper and start examining the following topics:

- Incompatible object collaboration with the adapter pattern
- Decoupling and developing objects independently with the bridge pattern
- Treating objects the same way using the composite pattern
- Extending object functionality by using the decorator pattern
- Simplifying communication with the facade pattern
- Using conditions to select desired objects with the filter pattern
- Sharing objects across an application with the flyweight pattern
- Handling requests with the front-controller pattern
- Identifying instances using the marker pattern
- Exploring the concept of modules with the module pattern
- Providing a placeholder for an object employing the proxy pattern
- Discovering multiple inheritance in Java with the twin pattern

By the end of this chapter, you'll have a solid understanding of how to structure a code base around created instances.

Technical requirements

You can find the code files for this chapter on GitHub at https://github.com/PacktPublishing/ Practical-Design-Patterns-for-Java-Developers/tree/main/Chapter04.

Incompatible object collaboration with the adapter pattern

The main goal of the adapter pattern is to connect the source class interface to another interface that clients will expect. The adapter pattern allows classes to work together that otherwise couldn't due to an incompatible abstraction or implementation. It is considered one of the most common patterns and is one of the **Gang of Four** (**GoF**) design patterns.

Motivation

The adapter pattern is also known as **wrapper**. An adapter wraps the behavior of the adaptee (connected class) and allows access to the adaptee without modification using an already existing interface. Commonly, an adaptee uses an incompatible interface, and an adapter consolidates such behavior and transparently provides access to the required functionality.

Finding it in the JDK

The java.base module provides multiple implementations of the adapter pattern. The Collections utility class from the java.util package provides a list method, which accepts an Enumeration interface and adapts the result into an ArrayList instance.

Sample code

The adapter pattern can be implemented in several ways. One of them is considered in the example of a vehicle that can have different types of engines (*Example 4.1*):

```
public static void main(String[] args) {
    System.out.println("Adapter Pattern: engines");
    var electricEngine = new ElectricEngine();
    var enginePetrol = new PetrolEngine();
    var vehicleElectric = new Vehicle(electricEngine);
    var vehiclePetrol = new Vehicle(enginePetrol);

    vehicleElectric.drive();
    vehicleElectric.refuel();
```

```
        vehiclePetrol.drive();
        vehiclePetrol.refuel();
}
```

Here is the output:

```
Adapter Pattern: engines
...
Vehicle, stop
Vehicle needs recharge
ElectricEngine, check plug
ElectricEngine, recharging
...
Vehicle needs petrol
PetrolEngine, tank
```

Example 4.1 – Although each type of vehicle shares similar logic, the behavior of the refuel method varies by engine type

These engines share some similar functionalities and features, but not all. They are very different from each other (*Figure 4.1*):

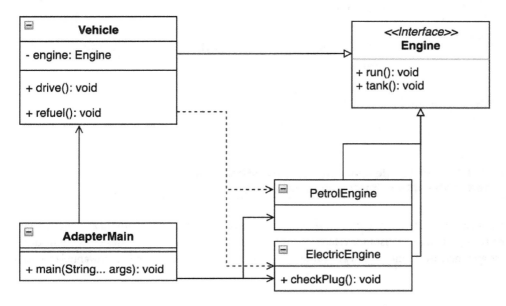

Figure 4.1 – UML class diagram highlighting engine type differences

In this example, the `Vehicle` class and its instance play the role of adapter. In the case of the `drive` method, both motors behave similarly. The `tank` method execution is a different scenario because the vehicle adapter needs to know the exact engine type to correctly execute the `refuel` method (*Example 4.2*):

```java
class Vehicle {
    private final Engine engine;


    ...


    void refuel(){
        System.out.println("Vehicle, stop");
        switch (engine){
            case ElectricEngine de -> {
                System.out.println("Vehicle needs diesel");
                de.checkPlug();
                de.tank();
            }
            case PetrolEngine pe -> {
                System.out.println("Vehicle needs petrol");
                pe.tank();
            }
            default -> throw new IllegalStateException
                ("Vehicle has no engine");
        }
        engine.tank();
    }
}
```

Example 4.2 – The Vehicle instance operates with an engine based on its type, identified by the pattern-matching functionality

New language features such as `switch` statement enhancements can be very useful as there is no need to keep the exact adaptee instance reference for the adapter. The `sealed` classes enforce the desired purpose and increase maintainability by protecting their intent, such as by avoiding unwanted extensions.

Both engine types considered may still implement similar abstraction in order to maintain the concept of the engine (*Example 4.3*):

```
sealed interface Engine permits ElectricEngine,
    PetrolEngine  {
    void run();
    void tank();
}
```

Example 4.3 – The Engine interface allows only certain classes to implement its methods

The `Vehicle` adapter provides the required logic to handle the different engine implementations properly. The `ElectricEngine` implementation provides an additional `checkPlug` method (*Example 4.4*):

```
final class ElectricEngine implements Engine{
    @Override
    public void run() {
        System.out.println("ElectricEngine, run");
    }


    @Override
    public void tank() {
        System.out.println("ElectricEngine, recharging");
    }


    public void checkPlug(){
        System.out.println("ElectricEngine, check plug");
    }
}
```

Example 4.4 – ElectricEngine implements additional logic that is not shareable with the general Engine concept

Conclusion

The adapter structural design pattern has a valid place in development as it represents a maintainable way to connect different functionalities and control them through a similar interface. The adapter is properly encapsulated and can be even more abstract. The new `sealed` classes support the pattern concept of maintainability and clarity. The consequence of using the adapter pattern may be that the adapter needs to commit to a specific adaptee or interface. The adapter may extend some of the adaptee functionalities as a subclass. The adapter pattern is worth considering when additional third-party libraries or APIs are to be implemented. It provides a transparent and decoupled way to interact with libraries, following the SOLID concept. Solutions can also be easily refactored.

This look at the adapter pattern has shown the approach of using incompatible APIs. Next, let us investigate how to use different replaceable implementations.

Decoupling and developing objects independently with the bridge pattern

The goal of this pattern is to separate the abstraction from its implementation so that both can change independently. The bridge pattern was described by the GoF.

Motivation

The bridge pattern is about prioritizing composition over inheritance. The implementation details are moved from the hierarchy to another object with a separate hierarchy. The bridge pattern uses encapsulation and aggregation, and may use inheritance to separate responsibilities into different classes.

Finding it in the JDK

Uses of the bridge pattern can be found in the `java.util.logging` package and the implementation of the `Logger` class. The class is located in the `java.logging` module. It implements the `Filter` interface. This interface is used to gain additional control over logged content beyond the standard log level.

Sample code

Let us see an example of two types of vehicles: a sport car and a pickup. The vehicles vary in engine type: petrol and diesel. The intention is to enforce a separate development for the `Vehicle` and `Engine` abstraction source code. The example case creates vehicles and executes `drive` and `stop` methods (*Example 4.5*):

```java
public static void main(String[] args) {
    System.out.println("Pattern Bridge, vehicle
```

```
        engines...");
    Vehicle sportVehicle = new SportVehicle(new
        PetrolEngine(), 911);
    Vehicle pickupVehicle = new PickupVehicle(new
        DieselEngine(), 300);

    sportVehicle.drive();
    sportVehicle.stop();

    pickupVehicle.drive();
    pickupVehicle.stop();
}
```

Here is the output:

```
Pattern Bridge, vehicle engines...
SportVehicle, starting engine
PetrolEngine, on
SportVehicle, engine started, hp:911
SportVehicle, stopping engine
PetrolEngine, self check
PetrolEngine, off
SportVehicle, engine stopped
PickupVehicle, starting engine
DieselEngine, on
PickupVehicle, engine started, hp:300
PickupVehicle, stopping engine
DieselEngine, off
PickupVehicle, engine stopped
```

Example 4.5 – The vehicles use different engines; they can be developed separately due to the bridge pattern's isolation

Each vehicle extends the Vehicle abstraction class that runs the engine and encapsulates the basic functions. The Engine interface, used by the vehicle abstraction, plays the role of a bridge, as shown in the following diagram (*Figure 4.2*):

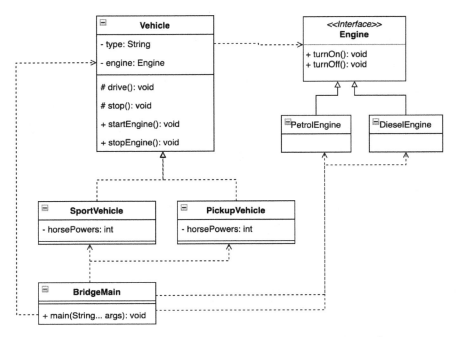

Figure 4.2 – The UML class diagram shows how the Engine interface
bridges access to a specific implementation

The engines already behave differently, and due to the bridge they may continue evolving (*Example 4.6*):

```java
class DieselEngine implements Engine{

        ...
    @Override
    public void turnOff() {...}
}
class PetrolEngine implements Engine{

        ...
    @Override
    public void turnOff() {
        selfCheck();

        ...
    }
    private void selfCheck(){ ...}
}
```

Example 4.6 – The engines differ in implementation

The vehicle abstraction does not have any engine implementation details, which may vary even in the class hierarchy. The vehicle only needs to rely on the provided interface.

Conclusion

The bridge pattern is a good idea to consider when the application source code requires reducing bindings to specific implementation classes. Due to the bridge pattern, the decision about a specific implementation can be deferred until runtime. The bridge pattern helps to encourage SOLID design principles through responsibility separation and encapsulation. The implementation can be freely tested and shared as required through the application source code. It is required to keep in mind not to add unwanted responsibilities to the bridge implementation and consider alternative approaches in terms of design patterns when such a situation takes a place.

The bridge pattern can open the door to better composition of implementation specifics, as we'll explore in the next pattern.

Treating objects the same way using the composite pattern

The composite pattern is a remarkable solution for handling objects uniformly while arranging them in a tree structure, which simplifies access to instances. The demand for it naturally came from industry, and the pattern was soon identified and described by the GoF.

Motivation

Grouping objects around the underlying business logic is a powerful approach. A composite design pattern outlines a way to achieve such a state. Since each member of the group is treated uniformly, it is possible to create hierarchical tree structures and part-whole hierarchies. It helps to establish the logical relationships of the application and the composition of the desired objects.

Finding it in the JDK

In the JDK, the composite pattern can be found in the `java.base` module, the `java.util` package, and the `Properties` class. The `Properties` class implements the `Map` interface through its `Hashtable` implementation, and also contains a `ConcurrentHashMap` instance to store the property values internally. Although the `Properties` class's `put` operation remains synchronized due to the `Hashtable` implementation, the `get` operation does not, as it is simple to read into the concurrent map.

Sample code

To explore the power of the composition pattern, consider the `SportVehicle` class, which implements the `Vehicle` interface. It is standard knowledge that every vehicle is a collection of parts and every part is a grouping of smaller parts (*Figure 4.3*):

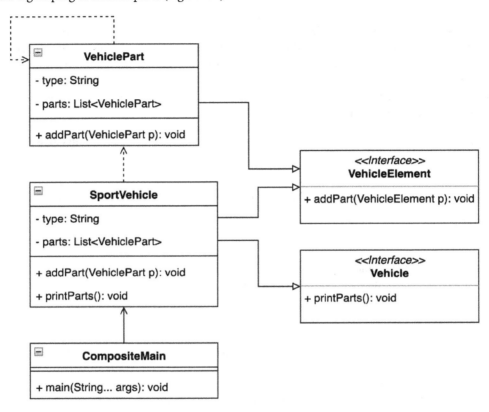

Figure 4.3 – The UML class diagram shows how SportVehicle is composed of VehiclePart types

When the vehicle manufacturing process starts, the composition pattern provides a complete collection of every part that is present in the final results (*Example 4.7*):

```
public static void main(String[] args) {
    System.out.println("Pattern Composite, vehicle
        parts...");
    var fastVehicle = new SportVehicle("sport");
    var engine = new VehiclePart("fast-engine");
    engine.addPart(new VehiclePart("cylinder-head"));
```

```
        var brakes = new VehiclePart("super-brakes");
        var transmission = new VehiclePart("automatic-
            transmission");
        fastVehicle.addPart(engine);
        fastVehicle.addPart(brakes);
        fastVehicle.addPart(transmission);
        fastVehicle.printParts();
}
```

Here's the output:

```
Pattern Composite, vehicle parts...
SportCar, type'sport', parts:'
[{type='fast-engine', parts=[{type='cylinder-head',
  parts=[]}]},
{type='super-brakes', parts=[]},
{type='automatic-transmission', parts=[]}]'
```

Example 4.7 – Reviewing the SportVehicle instance composition

Conclusion

The composite pattern allows the representation of the composition of a class in granular detail. It takes into account the smaller parts of the composite by creating part-whole hierarchies. While this provides advantages because each part is treated uniformly, it can lead to ignoring differences between parts. On the other hand, the composite pattern holds all the involved parts together in a transparent form.

Let us now examine how an individual object can be extended with additional functionality without changing the API.

Extending object functionality by using the decorator pattern

The decorator pattern provides the ability to add new functionality to objects by placing those objects in a decorator, Sothat a decorated instance provides extended functionality. The implementation of the decorator pattern is relatively simple and dynamic in languages such as Python and Kotlin. On the other hand, Java may provide more stability and maintainability of the source code through visibility and new enhancements, which can be very valuable. The decorator pattern was identified and described by the GoF.

Motivation

With the decorator pattern, you can dynamically attach additional responsibilities to an object. A decorator provides a flexible alternative to subclasses to extend the functionality of a class. The decorator can be added statically or dynamically without altering the current behavior of an object.

Finding it in the JDK

Uses of the decorator pattern can be found in the Java collections framework, the `java.base` module, and the `java.util` package. The `Collection` class contains different ways to use the decorator pattern. For example, the `unmodifiableCollection` method wraps the requested collection into an unmodifiable collection represented by an `UnmodifiableCollection` instance that acts as a decorator for the provided collection type, similar to other methods starting with `unmodifiable...` words. Another example is methods starting with the word `synchronized...` of the `Collections` utility class.

Sample code

When you think about the previous examples of vehicles, the decorator pattern can be considered a tuned vehicle. The standard `SportVehicle` class is like this. It implements the `Vehicle` interface to fulfill standard functions. The application designer later decides to improve the current state and creates a `TunedVehicleDecorator` class that wraps the standard vehicle, without having to change previous functions (*Figure 4.4*):

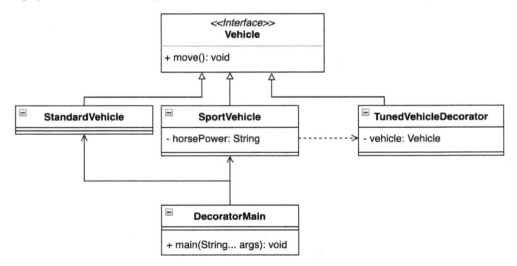

Figure 4.4 – The UML class diagram highlights the relationship between
SportVehicle and TunedVehicleDecorator types

All considered vehicles expose similar APIs to execute their implementations (*Example 4.8*):

```java
public static void main(String[] args) {
    System.out.println("Pattern Decorator, vehicle 1");
    Vehicle standardVehicle = new StandardVehicle();
    Vehicle vehicleToBeTuned = new StandardVehicle();
    Vehicle tunedVehicle = new SportVehicle
        (vehicleToBeTuned, 200);

    System.out.println("Drive a standard vehicle");
    standardVehicle.move();

    System.out.println("Drive a tuned vehicle");
    tunedVehicle.move();
}
```

Here is the output:

```
Pattern Decorator, tuned vehicle
Drive a standard vehicle
Vehicle, move
Drive a tuned vehicle
SportVehicle, activate horse power:200
TunedVehicleDecorator, turbo on
Vehicle, move
```

Example 4.8 – The tuned car abstraction extends the features of the SportVehicle type by adding more horsepower (200)

The decorator pattern may be introduced in multiple ways. In the presented example, TunedVehicleDecorator is an abstract class that holds the reference to the vehicle. The SportVehicle instance extends newly implemented functionality (*Example 4.9*):

```java
sealed abstract class TunedVehicleDecorator implements
    Vehicle permits SportVehicle {
    private final Vehicle vehicle;
    TunedVehicleDecorator(Vehicle vehicle) {
        this.vehicle = vehicle;
    }
```

```
        @Override
    public void move() {
        System.out.println("TunedVehicleDecorator,
            turbo on");
        vehicle.move();
    }
}


final class SportVehicle extends TunedVehicleDecorator {
    private final int horsePower;

    public SportVehicle(Vehicle vehicle, int horsePower) {
        super(vehicle);
        this.horsePower = horsePower;
    }

    @Override
    public void move() {
        System.out.println("SportVehicle, activate horse
            power:" + horsePower);
        super.move();
    }
}
```

Example 4.9 – Decorator wraps the Vehicle instance and extends its functionality

Conclusion

Class decoration can be very useful in many cases during application development. The decorator pattern can be used to migrate application logic where previous functionalities should remain hidden or unwanted subclassing should be avoided. The example showed how sealed classes can contribute to code maintainability and comprehensibility. Decoration helps not only to add new features but also to remove obsolete features. The decorator pattern is a transparent way to modify an object without disrupting the current interface.

Sometimes it makes sense to use the decorator pattern together with another design pattern we'll examine – the façade pattern.

Simplifying communication with the facade pattern

The facade pattern provides a unified interface to a set of underlying subsystems. In other words, a facade defines a higher-level interface that facilitates use. The facade pattern was described by the GoF.

Motivation

As subsystems evolve, they often become more complex. Most patterns, when used, result in smaller classes, thus making the subsystem more reusable and easier to customize, but also making it more difficult for all clients to work with. The facade pattern provides a simple default view of the subsystem that is good enough for most clients. Only clients who need more customizations will need to look beyond the façade pattern.

Finding it in the JDK

The Java collections framework resides in the `java.base` module and `java.util` has already been mentioned several times. It is a widely used part of the JDK, especially for internal logic implementation. Interfaces such as `List`, `Set`, `Queue`, `Map`, and `Enumeration` can be considered facades of a particular implementation. Let us review the `List` interface in more detail. It is implemented by the commonly used `ArrayList` or `LinkedList` classes and others. Implementation specifics vary in detail, some of which were mentioned in *Chapter 2, Discovering the Java Platform for Design Patterns* (*Tables 2.3, 2.4*, and *2.5*).

Sample code

The facade pattern is a frequently used design pattern in software engineering and is easily presented. Consider a case where a driver obtains their driving license for a vehicle. The driver's license entitles them to drive both gasoline and diesel cars and, of course, to refuel them. The driver gets both types as a reward (*Example 4.10*):

```java
public static void main(String[] args) {
    System.out.println("Pattern Facade, vehicle types");
    List<Vehicle> vehicles = Arrays.asList(new
        DieselVehicle(), new PetrolVehicle());
    for (var vehicle: vehicles){
        vehicle.start();
        vehicle.refuel();
    }
}
```

Here's the output:

```
Pattern Facade, vehicle types
DieselVehicle, engine warm up
DieselVehicle, engine start
DieselVehicle, refuel diesel
PetrolVehicle, engine start
PetrolVehicle, refuel petrol
```

Example 4.10 – The facade pattern promotes a standardized control interface

Consolidating vehicle types has a positive impact on code structure that is easy to implement (*Figure 4.5*):

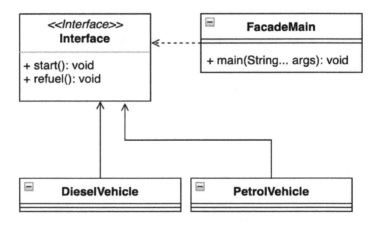

Figure 4.5 – The UML class diagram of the facade pattern usage for Vehicle implementations

Conclusion

The heavy use of the facade pattern makes it a good candidate to consider at any stage of application development. It promotes not only the principle of interface segregation but the entire SOLID concept. It helps implement internal dependencies while remaining customizable and maintainable. Facade helps introduce loose coupling and separates clients, forcing the removal of unwilling dependencies. The facade pattern naturally supports the horizontal scaling of the source code. Although the facade pattern provides a lot of benefits, misuse caused by unmaintained source code can turn into unwanted state. The solution is to re-evaluate the current implementation and apply improvements according to the SOLID principles.

Next, we'll examine how to select the correct object from a collection based on a rule.

Using conditions to select desired objects with the filter pattern

The filter pattern – sometimes called the criteria pattern – is a design pattern that allows clients to filter a set of objects using different criteria, or rules, and chain them separately using logical operations.

Motivation

The filter pattern helps simplify the code base to work like container objects that use subtyping instead of parameterization (generics) for an extensible class structure. It allows the client to easily extend and expose the filtering capability of container-like objects. Different filtering conditions can be dynamically added or removed without notifying the client.

Finding it in the JDK

Let us consider a filter as an interface with a single function and a logical Boolean result. A nice example of the filter pattern is the `Predicate` class, found in the `java.base` module and the `java.util.function` package. `Predicate` represents a Boolean function and is intended for use in the *Java Stream API* (discussed earlier, in *Chapter 2*, *Discovering the Java Platform for Design Patterns*), more specifically in the `filter` method, which accepts a predicate and returns a true or false primitive.

Sample code

A nice example of using the filter pattern would be an application that requires the selection of desired sensors in a vehicle. Every vehicle these days contains a huge number of sensors, so it can be difficult for a client to research each one individually (*Example 4.11*):

```
private static final List<Sensor> vehicleSensors = new
    ArrayList<>();
static {
    vehicleSensors.add(new Sensor("fuel", true));
    vehicleSensors.add(new Sensor("fuel", false));
    vehicleSensors.add(new Sensor("speed", false));
    vehicleSensors.add(new Sensor("speed", true));
}
public static void main(String[] args) {
    ...
    Rule analog = new RuleAnalog();
    Rule speedSensor = new RuleType("speed");
```

```
      ...

      var analogAndSpeedSensors = new RuleAnd(analog,
          speedSensor);
      var analogOrSpeedSensors = new RuleOr(analog,
          speedSensor);
      System.out.println("analogAndSpeedSensors=" +
          analogAndSpeedSensors.validateSensors
              (vehicleSensors));
      System.out.println("analogOrSpeedSensors=" +
              analogOrSpeedSensors.validateSensors
                  (vehicleSensors));
}
```

Here's the output:

```
Pattern Filter, vehicle sensors
AnalogSensors: [Sensor[type=fuel, analog=true],
    Sensor[type=speed, analog=true]]
SpeedSensors: [Sensor[type=speed, analog=false],
    Sensor[type=speed, analog=true]]
analogAndSpeedSensors=[Sensor[type=speed, analog=true]]
analogOrSpeedSensors=[Sensor[type=fuel, analog=true],
    Sensor[type=speed, analog=true], Sensor[type=speed,
        analog=false]]
```

Example 4.11 – Chaining a particular group of vehicle sensors with the filter pattern is simple and transparent

Let's draw an example (*Figure 4.6*):

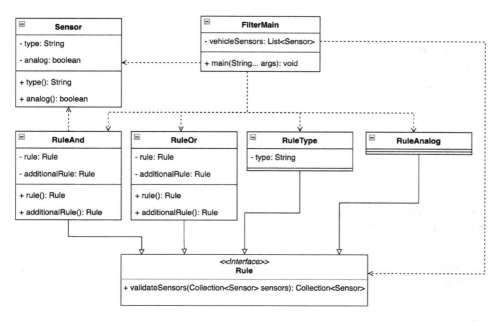

Figure 4.6 – The UML class diagram of possible rules used by the
container to select the proper Sensor instance

The Rule interface fulfills the expectation of the functional interface as it contains only one method, validateSensors. It also means that the compiler treats and optimizes the Rule interfaces like other annotated functional interfaces. Each rule can contain a specific implementation (*Example 4.12*):

```
@FunctionalInterface
interface Rule {
    Collection<Sensor> validateSensors(Collection<Sensor>
        sensors);
}
class RuleAnalog implements Rule {
    @Override
    public Collection<Sensor> validateSensors
        (Collection<Sensor> sensors) {
        return sensors.stream()
                .filter(Sensor::analog)
                .collect(Collectors.toList());
    }
```

```
    }
record RuleAnd(Rule rule, Rule additionalRule) implements
    Rule {
    @Override
    public Collection<Sensor> validateSensors
        (Collection<Sensor> sensors) {
        Collection<Sensor> initRule = rule.validateSensors
            (sensors);
        return additionalRule.validateSensors(initRule);
    }
}
```

Example 4.12 – Rules can contain trivial logic such as RuleAnalog, or advanced logic, such as RuleAnd, with respect to other rules running in the decision process

The sample application can be easily extended with any additional, more complex rule through a transparently defined interface.

Conclusion

Filtering or better selection of the correct instances may be required in places such as joining different request types or database results present in the Java heap. The filter pattern has shown its flexibility and that each rule can be developed independently, that is, optimized without the involvement of others, which makes it a suitable candidate when a client needs to work with container structures.

The next pattern represents a possible way to reduce the memory footprint by sharing instances.

Sharing objects across an application with the flyweight pattern

The flyweight pattern is used to minimize memory usage or computational cost by sharing as much as possible with similar objects. The flyweight pattern was described by the GoF author group.

Motivation

When a newly developed application uses many objects that are not required by the client. Memory maintenance costs can be high not only because of the large number of instances but also because of the creation of a new object. In many cases, such groups of objects can be successfully replaced by a relatively small number of instances. These instances can be transparently shared between the desired clients. This will reduce the pressure on the garbage collection algorithm. In addition, an application can reduce the number of open sockets when instances use such communication types.

Finding it in the JDK

The flyweight pattern can easily be found in the JDK. It may not be obvious to many. For example, in the implementation of primitive wrapper types, the java.base module and the java.lang package use this pattern to reduce memory overhead. A pattern is particularly useful when an application needs to handle many repeated values. Classes such as Integer, Byte, and Character provide a valueOf method, and its implementation uses an internal cache to store repeated elements.

Sample code

Let us examine an example case where a garage continually hires out specific vehicle types. The garage contains some vehicles that can be hired. Each has already prepared vehicle documents by default. When another vehicle is required, the new document is put into the system on demand (*Example 4.13*):

```
public static void main(String[] args) {
    System.out.println("Pattern Flyweight, sharing
        templates");
    Vehicle car1 = VehicleGarage.borrow("sport");
    car1.move();
    Vehicle car2 = VehicleGarage.borrow("sport");
    System.out.println("Similar template:" +
        (car1.equals(car2)));
}
```

Here's the output:

```
Pattern Flyweight, sharing vehicles
VehicleGarage, borrowed type:sport
Vehicle, type:'sport-car', confirmed
VehicleGarage, borrowed type:sport
Similar template: true
```

Example 4.13 – Sharing a template with the flyweight pattern is transparent and does not pollute memory

The heart of our next example (*Example 4.14*) is the implementation of VehicleGarage, which contains the cache for storing registration templates:

```
class VehicleGarage {
    private static final Map<String, Vehicle> vehicleByType
        = new HashMap<>();
```

```
    static {
        vehicleByType.put("common", new VehicleType
            ("common-car"));
        vehicleByType.put("sport", new VehicleType("sport-
            car"));
    }


    private VehicleGarage() {
    }

    static Vehicle borrow(String type){
        Vehicle v = vehicleByType.get(type);
        if(v == null){
            v =  new VehicleType(type);
            vehicleByType.put(type, v);
        }
        System.out.println("VehicleGarage, borrowed type:"
            + type);
        return v;
    }
}
```

Example 4.14 – The VehicleGarage implementation allows you to add a missing type on demand in order to control template size

The following example diagram shows that the client is not aware of the VehicleType class because it is not required (*Figure 4.7*):

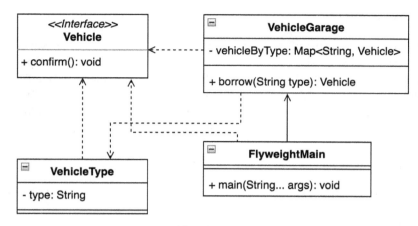

Figure 4.7 – The UML class diagram shows which classes VehicleGarage requires

Conclusion

The big advantage of the flyweight pattern is the ability to administrate a large number of requests for the required objects. It instantiates objects on demand and allows you to obtain control over the present instances. The application does not need to depend on the identity (`hashCode` and `equals`) of the object. The flyweight pattern provides a transparent way to obtain access to the object and its implementation enforces the SOLID design concept and DRY approach.

The next section describes how to consolidate incoming requests in a controlled manner.

Handling requests with the front-controller pattern

The goal of the pattern is to create a common service for most of the client requirements. The pattern defines a procedure that allows common functions such as authentication, security, custom manipulation, and logging to be encapsulated at a single location.

Motivation

This pattern is commonly seen within web applications. It implements and defines the standard handler used by the controller. It is the handler's responsibility to evaluate the validity of all incoming requests, although the handler itself may be available in many incarnations at runtime. The code is encapsulated in one place and referenced by the clients.

Finding it in the JDK

Usage of the front-controller pattern can be found in the `jdk.httpserver` module, the `sun.net.httpserver` package, and the `HttpServer` abstract class. The class implements the `createContext` abstract method, which accepts the `HttpHander` interface. Handler instances participate in HTTP request processing by executing the handler method. The release of JDK 18 comes with the `SimpleFileServer` wrapper of the underlying `HttpServer` implementations, available also as the standalone command `jwebserver` (*JEP-408: Simple Web Server*).

Sample code

Let us create a simple theoretical example not focused on parsing the web request (*Example 4.15*):

```
public static void main(String[] args) {
    System.out.println("Pattern FrontController, vehicle
        system");
    var vehicleController = new VehicleController();

    vehicleController.processRequest("engine");
    vehicleController.authorize();
    vehicleController.processRequest("engine");
    vehicleController.processRequest("brakes");
}
```

Here's the output:

```
Pattern FrontController, vehicle system
VehicleController, log:'engine'
VehicleController, is authorized
VehicleController, not authorized request:'engine'
VehicleController, authorization
VehicleController, log:'engine'
VehicleController, is authorized
EngineUnit, start
VehicleController, log:'brakes'
VehicleController, is authorized
BrakesUnit, activated
```

Example 4.15 – The vehicle system uses the front-controller pattern to process incoming commands

Imagine that vehicles contain a controller that is responsible for controlling brakes and motor units. All incoming commands are processed in this controller (*Figure 4.8*):

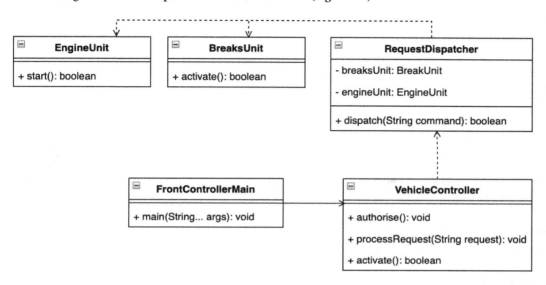

Figure 4.8 – The front-controller pattern enforces a loose coupling of the controller and dispatcher

The VehicleController object requires an instance of a specific handler. A handler is defined by an instance of the RequestDispatcher class (*Example 4.16*):

```
record RequestDispatcher(BrakesUnit brakesUnit, EngineUnit
    engineUnit) {
    void dispatch(String command) {
        switch (command.toLowerCase()) {
            case "engine" -> engineUnit.start();
            case "brakes" -> brakesUnit.activate();
            default -> throw new IllegalArgumentException
                ("not implemented:" + command);
        }
    }
}
class VehicleController {
    private final RequestDispatcher dispatcher;
    ...
    void processRequest(String request) {
```

```
        logRequest(request);
        if (isAuthorized()) {
            dispatcher.dispatch(request);
        } else {
            System.out.printf("""
                VehicleController, not authorized request:
                    '%s'%n""", request);
        }
    }
}
```

Example 4.16 – The request handler representation RequestDispatcher instance needs to be injected into VehicleController

Both the `BrakesUnit` and `EngineUnit` classes are separated from the handling or control logic and can be developed independently.

Conclusion

The main use of the front-controller pattern is in web frameworks, in addition to encapsulating requests for handling requests and increasing the portability of different types of handlers. These tools only need to be properly registered and run at runtime. Based on the implementation, the pattern supports dynamic handling behavior without the requirement to replace the class at runtime. The front-controller pattern introduces a centralized mechanism for processing incoming information.

Software design sometimes requires the dissemination of specific information for a group of classes. For such purposes, tagging is well worth considering. Let's dive deeper into it in the next section.

Identifying instances using the marker pattern

This pattern is extremely useful in identifying instances at runtime for specific treatment, such as triggering the desired action when an instance is available.

Motivation

The marker interface pattern represents an empty interface. Such an interface is used to identify a special group of classes at runtime. Because of this fact, the maker pattern is sometimes called tagging, as its sole purpose is to distinguish a special type of instance. The application thus provides the possibility to use special handling for such cases at runtime. Logic can be separated and properly encapsulated. Because annotation represents a special form of interface, Java implements the marker interface in two ways – a class can inherit from an interface or be annotated.

Finding it in the JDK

A clearer example of using the marker interface in the JDK can be found in the `java.base` module. The `java.io` package defines the `Serializable` interface and the `java.lang` package provides the `Cloneable` interface. Both do not implement any method, and both are used to inform the runtime about special handling. The `Serializable` interface is important during the serialization and deserialization processes (the `writeObject` and `readObject` methods), where each nested field requires an interface implementation to obtain the state of the instance while traversing the object graph. In a similar way, the `Cloneable` interface informs the JVM that the `Object.clone()` method is being used and it can create a field-to-field copy of the object. It is good to be aware of the field differences. The primitive types provide values but object-only references. It means that objects require an implementation of the `Cloneable` interface to provide a copy.

Sample code

Let us draw a simple real-world example where a vehicle contains multiple sensors (*Figure 4.9*):

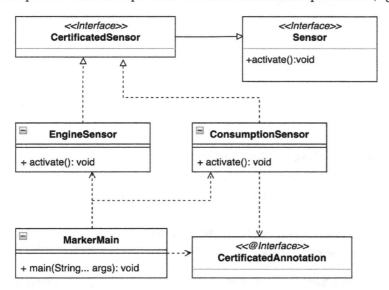

Figure 4.9 – UML class diagram of certified sensor tags (CertifiedSensor
and CertifiedAnnotation) using the marker interface pattern

The vehicle controller needs to identify special groups of the sensors that are certified for delivering specific information (*Example 4.17*):

```
public static void main(String[] args) {
    System.out.println("Pattern Marker, sensor
        identification");
```

```
        var sensors = Arrays
              .asList(new BrakesSensor(), new EngineSensor()
                    , new ConsumptionSensor());
        sensors.forEach(sensor -> {
            if(sensor.getClass().isAnnotationPresent
                (CertifiedAnnotation.class)){
                System.out.println("Sensor with Marker
                    annotation:" + sensor);
            } else {
                switch (sensor){
                    case CertifiedSensor cs -> System.out.
                        println("Sensor with Marker interface:
                            " + cs);
                    case Sensor s -> System.out.println
                        ("Sensor without identification:"+ s);
                }
            }
        });
    }
```

Here is the output:

```
Pattern Marker, sensor identification
Sensor without identification:BrakesSensor[]
Sensor with Marker interface:chapter04.marker
  .EngineSensor@776ec8df
Sensor with Marker annotation:chapter04.marker
  .ConsumptionSensor@30dae81
```

Example 4.17 – The marker interface pattern for sensor identification using the switch pattern-matching construct

This example introduced both types of pattern usage. It defines the CertifiedAnnotation and CertifiedSensor interface.

To group all sensor kinds during the implementation, the Sensor interface is used (*Example 4.18*):

```
@Retention(RetentionPolicy.RUNTIME)
@interface CertifiedAnnotation {}
```

```
public interface CertifiedSensor extends Sensor {}
public interface Sensor {
    void activate();
}
```

Example 4.18 – Implementation of the tagging interfaces CertifiedAnnotation and CertifiedSensor, and Sensor abstraction with methods

Using tags is trivial. A class must be annotated or inherit the marker interface (*Example 4.19*):

```
@CertifiedAnnotation
class ConsumptionSensor implements Sensor {
    @Override
    public void activate() {...}
}

final class EngineSensor implements CertifiedSensor {
    @Override
    public void activate() {...}
}
```

Example 4.19 – Marker usage for the sensor identification

Conclusion

The marker interface pattern can be a powerful tool at runtime, but it must be used wisely as it can have some drawbacks. One is that the purpose of using the marker pattern may be forgotten, or it may become obsolete as the application evolves. The second is the implementation of special handling logic. Distributing such logic can negatively affect application behavior. On the other hand, a marker interface can simplify application logic, and in many cases, an interface is preferred over an annotation because it is more traceable.

Let us introduce a vehicle unit's modularity in the next pattern.

Exploring the concept of modules with the module pattern

This pattern implements the concept of software modules defined by modular programming. The pattern is used in cases where the programming language does not have direct support for such a concept or the application requires it.

Motivation

This pattern can be implemented in several ways depending on the application requirements. The module pattern concentrates or encapsulates the composition of an application's functionality into precisely identified modules. The Java platform has already implemented basic support for the module concept through the Jigsaw project, available since the release of JDK 9, but it is possible to try to create it programmatically in a similar way, although not entirely in isolation, as the source code can influence its modularization approach.

Finding it in the JDK

The best example that can be found in the JDK of the module pattern is the Java platform modules. This concept was discussed in great detail in *Chapter 2, Discovering the Java Platform for Design Patterns*, in the *Getting to grips with the Java Module System* section.

Sample code

Let us imagine a vehicle that need to have isolated brakes and engine systems. This is pretty much according to a real-world scenario. Each module will operate independently, and only one provider is present at runtime. Before the vehicle can be used, both modules need to be activated (*Example 4.20*):

```java
class ModuleMain {
    ...

    private static void initModules() {
        brakesModule = BrakesModule.getInstance();
        engineModule = EngineModule.getInstance();
        engineModule.init();
    }

    ...

    public static void main(String[] args) {
        initModules();
        printStatus();
    }
}
```

Here's the output:

```
BrakesModule, unit:BrakesModule@5ca881b5
EngineModule, unit:EngineModule@4517d9a3
```

```
EngineModule, init
BrakesModule, ready:false
EngineModule, ready:true
```

Example 4.20 – The client function initModules activates the modules correctly in encapsulation

The following diagram emphasizes the separation of modules, although the programmatic approach allows sharing or implementing shared abstractions (*Figure 4.10*):

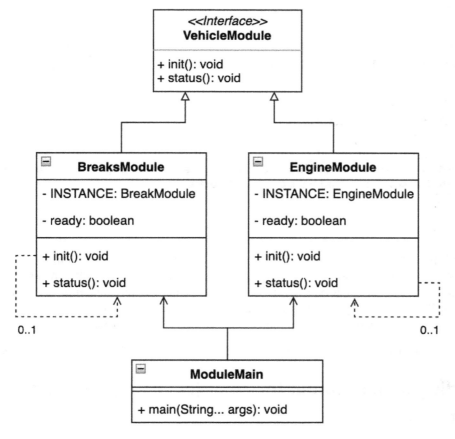

Figure 4.10 – UML class diagram showing a pattern implementation
realized by the provided VehicleModule interface

Each module is represented as a singleton instance in order to ensure only one instance provides a transparent gateway to access the module functionality:

```
class EngineModule implements VehicleModule {
    private static volatile EngineModule INSTANCE;
    static EngineModule getInstance() {

        ...

        return INSTANCE;
    }
    private boolean ready;

    ...

    @Override
    public void init() {...}

    @Override
    public void status() {...}
}
```

Example 4.21 – The EngineModule and BrakesModule example implementations are represented by singletons and have a similar structure

Conclusion

The module pattern introduces structure to the source code in a very transparent way. Each module can be developed independently without influence. Because a programmatic solution may not fully enforce source code isolation, it is necessary to extend modules wisely. Another drawback may be module initialization, as a singleton may not be an acceptable solution. On the other hand, the module pattern provides a workflow to develop a source code with all SOLID concepts in mind.

What about using proxies instead of modules and implementations? Let us dive deeper in the next section.

Providing a placeholder for an object using the proxy pattern

The proxy pattern is considered a placeholder that manages access to another object in order to gain control of it. The pattern may also be known by the name surrogate. The proxy pattern was described by the GoF.

Motivation

In its most general form, a proxy is a class acting as an interface to the client. A proxy is considered a wrapper or agent object that is used by a client. The client accesses the actual object through the same interface and the actual implementation stays hidden from the client in the background. Communication between the client and the implementation remains transparent, thanks to the proxy pattern.

By using a proxy, the client can access the actual object, or it can provide additional logic.

Finding it in the JDK

The proxy design pattern also has a place in the JDK. The most well-known one is the public `Proxy` class, which you can find in the `java.reflect` package of the `java.base` module. The `Proxy` class provides several static methods for creating objects used for method invocation.

Sample code

The given example can be considered as the remote control of a vehicle. A controller, represented by a proxy design pattern, provides exactly the same functionality as a real vehicle, also managing the connection between the real vehicle instance (*Example 4.22*):

```
public static void main(String[] args) {
    System.out.println("Pattern Proxy, remote vehicle
        controller");
    Vehicle vehicle = new VehicleProxy();
    vehicle.move();
    vehicle.move();
}
```

Here's the output:

```
Pattern Proxy, remote vehicle controller
VehicleProxy, real vehicle connected
VehicleReal, move
VehicleReal, move
```

Example 4.22 – The VehicleProxy instance works like a real vehicle

The real vehicle implementation is defined by the generic abstraction provided by the `Vehicle` interfaces (*Figure 4.11*):

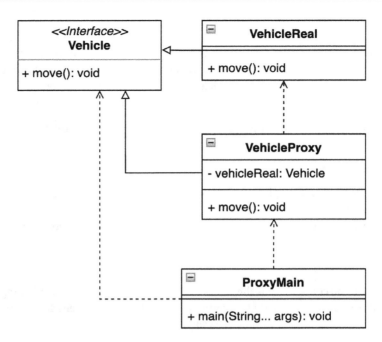

Figure 4.11 – An example of a vehicle proxy can be depicted by a UML class diagram

This allows seamlessly extending the types of controlled vehicles, as shown in the proxy pattern implementation (*Example 4.23*):

```
class VehicleProxy implements Vehicle{
    private Vehicle vehicleReal;
    @Override
    public void move() {
        if(vehicleReal == null){
            System.out.println("VehicleProxy, real vehicle
                connected");
            vehicleReal = new VehicleReal();
        }
        vehicleReal.move();
    }
}
```

Example 4.23 – The VehicleProxy class contains a reference to the actual Vehicle instance

Conclusion

The proxy design pattern brings many advantages to the source code, for example, the implementation can be replaced at runtime. In addition to being used to fully control access to the actual instance, it can also be used for lazy initiation, as we saw in *Example 4.23*. The proxy has its legitimate place in driver implementation or network connections as it naturally enforces not only logging possibilities but also code separation through the segregation of the interfaces and other SOLID principles. It is useful to consider when an application requires I/O operations.

Java as a language does not support multiple inheritance, but it is still possible to achieve. Let us examine how in the next section.

Discovering multiple inheritance in Java with the twin pattern

This pattern allows you to combine functions of objects that tend to be used together, which is a common paradigm used by languages without multiple inheritance support.

Motivation

The twin pattern presents the possibility to implement multiple inheritance in Java. Multiple inheritance is not a supported concept as it may lead to compiler inconsistency, known as the diamond problem. The diamond problem defines a state through class abstraction where the compiler may turn out to be inconsistent. This state is due to the lack of information due to multiple abstract classes. The compiler does not have enough information about which methods should execute.

Sample code

This pattern is not supported by the platform and is rarely required for development. For these reasons, the pattern most likely does not exist inside the released JDK, as described. However, let us examine a possible example to better understand the pattern. Imagine the vehicle initiation sequence. During initiation, the engine and brake units need to be initiated together. In other words, when the engine is initiated, the brakes must be initiated too, and the other way around (*Example 4.24*):

```java
public static void main(String[] args) {
        System.out.println("Pattern Twin, vehicle
            initiation sequence");

        var vehicleBrakes1  = new VehicleBrakes();
        var vehicleEngine1 = new VehicleEngine();
        vehicleBrakes1.setEngine(vehicleEngine1);
```

```
        vehicleEngine1.setBrakes(vehicleBrakes1);

        vehicleEngine1.init();
    }
```

Here's the output:

```
Pattern Twin, vehicle initiation sequence
AbstractVehiclePart, constructor
AbstractVehiclePart, constructor
VehicleBrakes, initiated
VehicleEngine, initiated
```

Example 4.24 – The twin pattern gives a guarantee that both units are always initiated

The following diagram shows us tight coupling between units:

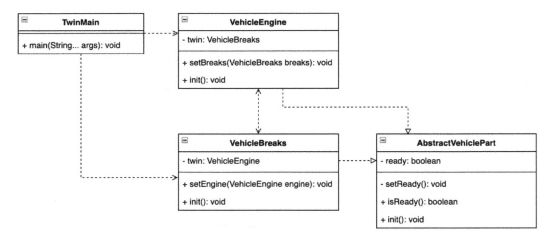

Figure 4.12 – Both considered units, VehicleEngine and VehicleBrakes, are very closely coupled

The coupling also translates into a code base that can be very fragile for future development (*Example 4.25*):

```
public class VehicleBrakes extends AbstractVehiclePart {

    private VehicleEngine twin;

    VehicleBrakes() {
```

```
    }

    void setEngine(VehicleEngine engine) {
        this.twin = engine;
    }


    @Override
    void init() {
        if (twin.isReady()) {
            setReady();
        } else {
            setReady();
            twin.init();
        }
        System.out.println("VehicleBrakes, initiated");
    }
}
```

Example 4.25 – The VehicleBrakes class implementation shows a tight coupling with its twin, VehicleEngine

Conclusion

The twin pattern can be used to achieve multiple inheritance in Java. It must be used wisely, as a logical unwritten requirement is to guarantee complete separation of the objects under consideration. In other words, the twin design pattern allows twins to function as a single instance with extended functionality and features.

Summary

Knowledge of structural patterns along with newly added Java syntax enhancements not only improves maintainability but also enforces all previously learned OOP concepts and improves responsiveness to potential deviations in code behavior such as exceptions, unexpected crashes, or logical issues.

We built a solid foundation through the examples in this chapter and learned how to use the adapter pattern to create collaboration between mutually incompatible objects, and also how to transparently separate an object's implementation from its abstraction using the bridge pattern. The composite pattern presented a way to organize and wrap objects into a tree structure around the underlying business logic. We investigated the possibility of expanding an object's functionality by using the decorator pattern. A way to simplify communication between objects was presented by the facade pattern, followed by the filter pattern, which allows us to select only the instances we want. We learned how the flyweight design pattern allows us to re-use already created runtime instances, and the approach of processing the incoming information was presented by the front-controller pattern, so that the client can only respond to valid requests. We discovered how the marker pattern allows a client to handle a specific group of objects in a unique way. We explored the possibility of modularizing the code base by implementing the module pattern. We saw how to use the proxy pattern to let a client indirectly gain control of an object without being aware of its implementation details, and we saw how to use the twin pattern to implement multiple inheritance in Java even though the language does not support it.

With the knowledge gained about creational and structural design patterns, the underlying source code structure is well organized and open to continuous application development. The next chapter examines the behavioral design patterns that help organize communication and responsibilities between targeted instances.

Questions

1. What challenges do structural design patterns solve?

2. Which structural design patterns are described by the Gang of Four?

3. Which design pattern is appropriate for creating a tree structure of related objects?

4. Which structural design pattern can be used to identify an object at runtime?

5. Which design pattern can be used for indirect object access, with the same functionality of the object itself?

6. Which design pattern promotes the separation of logic from its abstraction?

Further reading

- *Design Patterns: Elements of Reusable Object-Oriented Software* by Erich Gamma, Richard Helm, Ralph Johnson, and John Vlissides, Addison-Wesley, 1995

- *Design Principles and Design Patterns* by Robert C. Martin, Object Mentor, 2000

- *JSR-376: Java Platform Module System*, https://openjdk.java.net/projects/jigsaw/spec/

- *JSR-408: Simple Web Server*, https://openjdk.org/jeps/408

- *Clean Code* by Robert C. Martin, Pearson Education, Inc, 2009

- *Effective Java – Third edition* by Joshua Bloch, Addison-Wesley, 2018

- *Twin – A Design Pattern for Modelling Multiple Inheritance*, Hanspeter Mössenböck, University of Linz, Institute for System Software, 1999, https://ssw.jku.at/Research/Papers/Moe99/Paper.pdf

5

Behavioral Design Patterns

Code maintainability plays a key role in applications across the spectrum of the industry, but it's not fair to stop there and look no further. This means skipping over code behavior at runtime, which has an impact on physical and virtual memory usage. The primary motivation for using behavior patterns is transparent communication between objects, or in other words, the efficient usage of memory allocation for this communication. Utilizing behavioral patterns improves the flexibility of communication and helps complete a task with a single object or multiple objects exchanging information with each other. Structural design patterns can sometimes seem close to behavioral ones, but as we will see in each case, the purpose is slightly different. Let us dive deeper and learn about the following:

- Limiting expensive initialization using the caching pattern
- Handling events using the chain of responsibility pattern
- Turning information into action with the command pattern
- Giving meaning to the context using an interpreter pattern
- Checking all the elements with the iterator pattern
- Utilizing the mediator pattern for information exchange
- Restoring the desired state with the memento pattern
- Avoiding a null pointer exception state with the null object pattern
- Keeping all interested parties informed using the observer pattern
- Dealing with instance stages by using the pipeline pattern
- Changing object behavior with the state pattern
- Using the strategy pattern to change object behavior
- Standardizing processes with the template pattern
- Executing code based on the object type using the visitor pattern

By the end of this chapter, you will have a good foundation for understanding the importance of program behavior, not only for resource utilization but also from the perspective of SOLID design principles.

Technical requirements

You can find the code files for this chapter on GitHub at `https://github.com/PacktPublishing/Practical-Design-Patterns-for-Java-Developers/tree/main/Chapter05`.

Limiting expensive initialization using the caching pattern

The caching pattern is not found in the traditional list from the **Gang of Four** (**GoF**). However, due to industry requirements and resource usage, it has been identified as a commonly used approach and has gained importance.

Motivation

The caching pattern supports element reuse. It does not create a new element on demand – instead, it reuses an already-created element stored in the cache. It stores frequently needed data in fast-access storage for increased performance. Reading data from the cache is faster than instantiating a new entity given the low complexity of fetching the required element.

Finding it in the JDK

The `java.base` module and its `java.lang` package provide wrapper classes for primitive types. The `valueOf` method for double, float, integer, byte, or character types uses a caching mechanism for frequently requested values to reduce memory space and improve performance.

Sample code

Let us imagine the following caching example by creating a `Vehicle` class. It contains references to its internal systems (*Figure 5.1*):

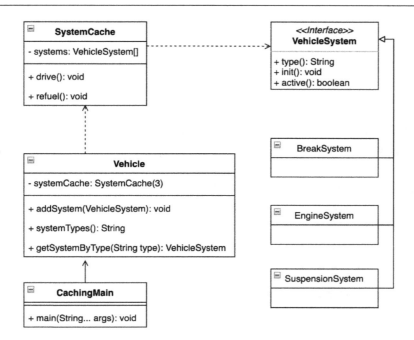

Figure 5.1 – The UML class diagram showing the considered VehicleSystem types for the Vehicle class

This means that the vehicle is precisely defined and does not mutate. When a client requests a specific system, it always initiates the one that corresponds to the object. This pattern also enforces control over the storage procedure (*Example 5.1*):

```java
public static void main(String[] args) {
    System.out.println("Caching Pattern, initiated vehicle
        system");
    var vehicle = new Vehicle();
    vehicle.init();
    var suspension = new SuspensionSystem("suspension");
    vehicle.addSystem(suspension);
    System.out.printf("Systems types:'%s%n",
        vehicle.systemTypes());

    var suspensionCache =
        vehicle.getSystemByType("suspension");
    System.out.printf("Is suspension equal? '%s:%s'%n",
        suspension.equals(suspensionCache),
```

```
                suspensionCache);
    vehicle.addSystem(new EngineSystem("engine2"));
}
```

Here's the output:

```
Caching Pattern, initiated vehicle system
Vehicle, init cache:'break':'BreakSystem@adb0cf77',
  'engine':'EngineSystem@a0675694'
Systems types:''break':'BreakSystem@adb0cf77','engine'
  :'EngineSystem@a0675694','suspension':'Suspension
    System@369ef459'
Is suspension equal? 'true:SuspensionSystem@369ef459'
SystemCache, not stored:EngineSystem@6c828066
```

Example 5.1 – The caching pattern provides a guarantee to always get the required element and gain control over the storage

In other words, it is not necessary to create another instance such as EngineSystem to access its functions. Those object accesses or that program behavior can easily lead to undesirable conditions.

The vehicle's SystemCache only considers specific types of instances and is also limited by size (*Example 5.2*):

```
class SystemCache {
    private final VehicleSystem[] systems;
    private int end;

    ...

    boolean addSystem(VehicleSystem system) {
        var availableSystem = getSystem(system.type());
        if (availableSystem == null && end <
            systems.length) {
            systems[end++] = system;
            return true;
        }
        return false;
    }
```

```
    VehicleSystem getSystem(String type)  {...}


    ...

}
```

Example 5.2 – The SystemCache instance provides features to ensure program stability and may give additional guarantees

Conclusion

The example (from *Figure 5.1*) showed that implementing caching is simple. It may be a good idea to consider when clients require repeated access to the same set of elements. This can have a positive impact on performance.

Some of these elements may be responsible for the program's runtime behavior. Let's dive deeper into this in the next section.

Handling events using the chain of responsibility pattern

The chain of responsibility pattern helps avoid tying the handler logic to the sender that fired the event. This pattern was identified by the GoF's book.

Motivation

The program receives an initial triggered event. Each chained handler decides whether to process the request or pass it on to the next handler without responding. A pattern can consist of command objects that are processed by a series of handler objects. Some handlers can act as dispatchers, capable of sending commands in different directions to form a responsibility tree.

The chain of responsibility pattern allows you to build a chain of implementations in which a certain action is performed before or after calling the next handler in the chain.

Finding it in the JDK

The `java.logging` module includes the `java.util.logging` package, which contains a `Logger` class, intended for recording application component messages. Loggers can be chained and the logged message is only processed by the desired `Logger` instances.

Another example provided in the JDK is the `DirectoryStream` class, which comes with the `java.base` module and is located in the `java.nio` package. This class is responsible for iterating over entire directories and contains a nested filter interface. The interface provides an `accept` method. The actual representation of the chained filter differentiates depending on whether the directory is to be processed or excluded.

Sample code

Let us examine an example of how the chain of responsibility design pattern can be used to respond to a trigger event from the driver system (*Example 5.3*):

```
System.out.println("Pattern Chain of Responsibility, vehicle
    system initialisation");
var engineSystem = new EngineSystem();
var driverSystem = new DriverSystem();
var transmissionSystem = new TransmissionSystem();


driverSystem.setNext(transmissionSystem);
transmissionSystem.setNext(engineSystem);


driverSystem.powerOn();
}
```

Here's the output:

```
Pattern Chain of Responsibility, vehicle system initialisation
DriverSystem: activated
TransmissionSystem: activated
EngineSystem, activated
```

Example 5.3 – The DriverSystem instance initiates the powerOn event that is propagated through the chained instances

The behavior of the system chain created is transparent, and the logic is properly encapsulated by each system. The provided generic abstract `VehicleSystem` instance defines the functionality, what function each element must fulfill, and how the following elements should be chained (*Example 5.4*):

```
sealed abstract class VehicleSystem permits DriverSystem,
    EngineSystem, TransmissionSystem {
    ...
    protected VehicleSystem nextSystem;
    protected boolean active;


        ...
    void setNext(VehicleSystem system){
        this.nextSystem = system;
```

```
        }

    void powerOn(){
        if(!this.active){
            activate();
        }
        if(nextSystem != null){
            nextSystem.powerOn();
        }
    }
}
```

Example 5.4 – The sealed classes usage provides additional stability and control

The client receives a framework for how and which classes can be considered when building a chain (*Figure 5.2*):

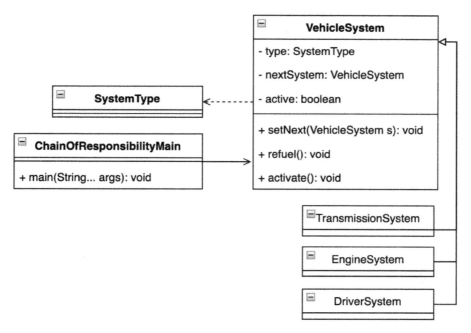

Figure 5.2 – The UML class diagram showing which elements participate in the powerOn event

Conclusion

The chain of responsibility pattern showed that an incoming event that affects the runtime behavior of a program can result in the creation of multiple objects. The manipulators are encapsulated and the logic is properly isolated according to SOLID principles. Using this pattern, the client gets the opportunity to dynamically decide which handlers should be involved in the event process. Therefore, it is a hot candidate for security frameworks or similar.

Chained handlers can issue multiple commands to the client at runtime. Let's explore command responses in more detail.

Turning information into action with the command pattern

The command pattern is sometimes known as action. The command pattern encapsulates the triggered event as an object that allows the client to act. This pattern was early identified and described in the GoF's book.

Motivation

The command pattern dictates which instances of the command interface perform which actions on the receiver client. A command object can be parameterized to define an action in greater detail. The commands can include a callback function to notify others of the occurrence of an event. Sometimes, commands can be thought of as object-oriented replacements for callback functions. A newly created command object can have different dynamics depending on the event that initiated it. The client can react to it according to an already scheduled scenario.

Finding it in the JDK

Nice examples are provided in the JDK by the `Callable` and `Runnable` interfaces from the `java.base` module and the `java.util.concurrent` package. The implementation of each interface is scheduled for execution based on a known scenario.

Other uses of the command pattern can be found in the `java.desktop` module in the `javax.swing` package and a class implementing the `Action` interface.

Sample code

The following example shows how a `Driver` object controls a vehicle using well-defined commands (*Example 5.5*):

```
public static void main(String[] args) {
    System.out.println("Pattern Command, turn on/off
        vehicle");
```

```
    var vehicle = new Vehicle("sport-car");
    var driver = new Driver();
    driver.addCommand(new StartCommand(vehicle));
    driver.addCommand(new StopCommand(vehicle));
    driver.addCommand(new StartCommand(vehicle));

    driver.executeCommands("start_stop");
}
```

Here's the output:

```
Pattern Command, turn on/off vehicle
START:Vehicle{type='sport-car', running=true}
STOP:Vehicle{type='sport-car', running=false}
START:Vehicle{type='sport-car', running=true}
```

Example 5.5 – The triggered start_stop event is translated and transformed by the Driver instance into actions

Commands are properly encapsulated and may contain additional logic for interacting with different clients or may decide on execution steps (*Example 5.6*):

```
sealed interface VehicleCommand permits StartCommand,
    StopCommand {
    void process(String command);
}
record StartCommand(Vehicle vehicle) implements
    VehicleCommand {
    @Override
    public void process(String command) {
        if(command.contains("start")){ ... }
}
```

Example 5.6 – To preserve the intended command design, the concept of sealed classes may be appropriated

Commands can be transparently extended as the driver functionality evolves (*Figure 5.3*):

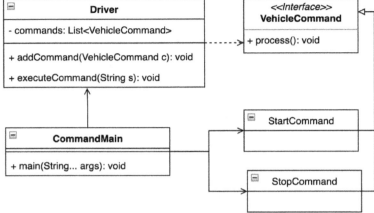

Figure 5.3 – The UML class diagram showing which command can be considered a driver

Conclusion

A simple example (*Figure 5.3*) showed the value of the command pattern. The command object is separate from the logic and may contain additional valuable information. A command has its own lifecycle and makes it easy to implement callback functions that can trigger another event.

Representing these commands textually can be tricky. The following section shows how the client can understand them.

Giving meaning to the context using the interpreter pattern

The interpreter pattern interprets sequences of characters into desired actions. It was identified early due to its use in SQL statement translation and described in more detail in the GoF's book.

Motivation

The interpreter pattern defines two types of objects that refer to specific sequences of characters. They are terminal and non-terminal actions or operations that can be performed on the sequence of characters under consideration. These operations represent the computer language that is used and have their own semantics. The syntactic tree of a given sentence – a sequence of characters – is an instance of a compound pattern and is used to evaluate and interpret meaning for the client program.

Finding it in the JDK

The `java.base` module contains the `java.util.regex` package with the `Pattern` class. This class represents the compilation of regular expressions. Specific semantics are applied to a sequence of characters to verify the required match.

Another example comes from a similar module and the `java.text` package. The abstract `Format` class implementation serves locale-sensitive information such as dates, number formats, and so on.

Sample code

Let's create a simple string math formula. The formula contains values from different sensors and their contribution to the result. The result represents the final value of the formula (*Example 5.7*):

```
public static void main(String[] args) {
    System.out.println("Pattern Interpreter, sensors
        value");
    var stack = new Stack<Expression>();
    var formula = "1 - 3 + 100 + 1";
    var parsedFormula = formula.split(" ");

    var index = 0;
    while (index < parsedFormula.length ){
        var text = parsedFormula[index++];
        if(isOperator(text)){
            var leftExp = stack.pop();
            var rightText = parsedFormula[index++];
            var rightEpx = new IntegerExpression
                (rightText);
            var operatorExp = getEvaluationExpression(text,
                left, right);
            stack.push(operatorExp);
        } else {
            var exp = new IntegerExpression(text);
            stack.push(exp);
        }
    }
    System.out.println("Formula result: " +
        stack.pop().interpret());
}
```

Here's the output:

```
Pattern Interpreter, math formula evaluation
Formula result: 99
```

Example 5.7 – The parser converts the math string formula into the appropriate Expression types

The basic element is the interface and its `interpret` method. The `1 - 3 + 100 + 1` formula is evaluated sequentially, and the last element contains the result. Each expression is encapsulated, and the interpreter can be conveniently extended (*Figure 5.4*):

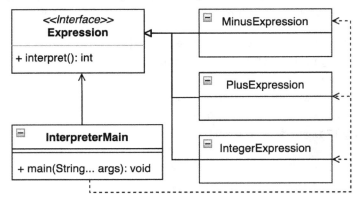

Figure 5.4 – The UML class diagram showing which participants are required for evaluation

Conclusion

An interpreter pattern is a powerful design pattern in which a client requires a textual representation of commands with known semantics to be processed. It allows you to create a grammar represented by a hierarchy of classes that can be easily extended and even dynamically modified at runtime.

The next section shows us how to navigate through a collection of commands. Let's dive into it.

Checking all the elements with the iterator pattern

This iterator pattern can be close to the abstraction of a cursor pointing to the desired position. Since array construction is a commonly used structure type, this pattern was soon identified. It is considered one of the core patterns contained in the GoF's book.

Motivation

The iterator pattern defines a transparent way to iterate through a collection of objects without having to expose or be aware of any of the internal details of an object. To step between elements, the iterator pattern uses an iteration function.

Finding it in the JDK

The java.base module contains multiple implementations of the iterator pattern. The first implementation can be considered the one provided by the collection framework located in the java. util package. An implementation of the Iterator interface traverses through a collection without knowing about the element's membership.

Another example that can be considered is the iterator provided by the BaseStream interface and its iterator method. This class comes from a similar module and the java.util.stream package, namely the Stream API. It represents the terminal operation.

Sample code

Every vehicle has several common parts. The following example shows the use of the iterator pattern to loop through them (*Example 5.8*):

```
public static void main(String[] args) {
    System.out.println("Iterator Pattern, vehicle parts");
    var standardVehicle = new StandardVehicle();
    for(PartsIterator part = standardVehicle.getParts();
        part.hasNext();) {
        var vehiclePart = part.next();
        System.out.println("VehiclePart name:" +
            vehiclePart.name());
    }
}
```

Here's the output:

```
Iterator Pattern, vehicle parts
VehiclePart name:engine
VehiclePart name:breaks
VehiclePart name:navigation
```

Example 5.8 – To preserve the intended command design, the concept of sealed classes may be appropriated

The vehicle can provide a common abstraction vehicle that handles the iterator (*Example 5.9*):

```
interface PartsIterator {
    boolean hasNext();
    VehiclePart next();
}
```

Example 5.9 – A program can implement different iterators with different dynamics

The client can traverse all elements individually. An implementation of this kind of iterator can be thought of as a nested class of a concrete implementation (*Example 5.10*):

```
sealed interface Vehicle permits StandardVehicle {
    PartsIterator getParts();
}


final class StandardVehicle implements Vehicle {

    private final String[] vehiclePartsNames = {"engine",
        "breaks", "navigation"};

    private class VehiclePartsIterator implements
        PartsIterator {


        ...

    }

    @Override
    public PartsIterator getParts() {
        return new VehiclePartsIterator();
    }
}
```

Example 5.10 – A program can implement different iterators with different dynamics

The sample program behavior is transparent to the client, and it provides a framework for extending the intended implementation of the vehicle with new parts and a way to navigate through them (*Figure 5.5*):

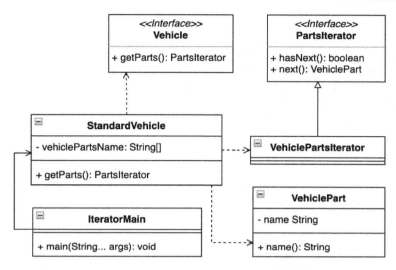

Figure 5.5 – The UML class diagram representing a collection of parts designed for StandardVehicle

Conclusion

The strength of the iterator pattern is that it can be implemented in a very general way – no need to understand the element types being considered. The iterator loops through them all without touching their internal representation at runtime. Along with another pattern, it can change strategies on the fly or only consider specific object types.

The next section explores runtime communication between specific object types – let's rock it.

Utilizing the mediator pattern for information exchange

A common situation across different types of applications is the requirement to manage communication between clients that require the exchange of information in order to maintain a process. This pattern was identified early and is one of the core patterns of the GoF's book.

Motivation

The mediator pattern represents an object, a man in the middle, that defines the way that a group of objects interacts within the group. The mediator establishes a free connection for client communication. Clients can refer to each other explicitly through an intermediary. In this way, communication can be moderated.

Finding it in the JDK

Although it may not be obvious at first glance, the mediator pattern can be easily found in the `java.base` module and the `java.util.concurrent` package. The `ExecutorService` class defines a `submit` method. Its parent class, `Executor`, exposes the `execute` method. These methods can be used to pass implementations of the `Callable` or `Runnable` interfaces, which were previously referred to as implementations of the command pattern.

Sample code

The following example is quite trivial compared to others, but it shows how a processor can maintain vehicle sensor communication (*Example 5.11*):

```java
record Sensor(String name) {
    void emitMessage(String message) {
        VehicleProcessor.acceptMessage(name, message);
    }
}
public static void main(String[] args) {
    System.out.println("Mediator Pattern, vehicle parts");
    var engineSensor = new Sensor("engine");
    var breakSensor = new Sensor("break");

    engineSensor.emitMessage("turn on");
    breakSensor.emitMessage("init");

}
```

Here's the output:

```
Mediator Pattern, vehicle parts
Sensor:'engine', delivered message:'turn on'
Sensor:'break', delivered message:'init'
```

Example 5.11 – Communication is handled by a VehicleProcessor instance

The central element of the example is the `VehicleProcessor` instance, which obtains all the messages emitted and can react to them (*Figure 5.6*):

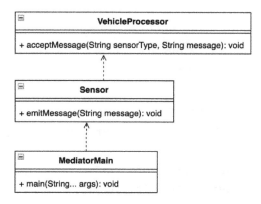

Figure 5.6 – The UML class diagram emphasizing the communication that takes place via the processor

Conclusion

The mediator pattern introduces the ability to isolate complex communication between different objects. The number of participating objects may vary at runtime. The pattern provides an encapsulated and decoupled way to allow all clients to communicate with each other.

Communication can bring about various states. In the next section, we'll explore how to remember and restore them.

Restoring the desired state with the memento pattern

Sometimes, it may be useful to consider keeping minimal information about the state of an object in order to continue or restore it. The memento pattern provides this functionality and was described in the GoF's book.

Motivation

Without breaking encapsulation, the internal state of an object, a memento, needs to be captured and externalized so that the object can later be restored to that state. The memento pattern provides a client function to restore the desired state of an object, a memento, on demand.

Finding it in the JDK

The Date class that comes with the java.base module and its java.util package is a nice implementation of the memento pattern. An instance of a class represents a specific point on a timeline, and the date can be restored to that timeline referring to the calendar or zone used.

Sample code

Let us look at the example of air conditioning in a vehicle. The controller gives us several options for setting the cockpit temperature, which also means that the driver can restore an already selected state (*Example 5.12*):

```java
public static void main(String[] args) {
    System.out.println("Memento Pattern, air-condition
        system");
    var originator = new AirConditionSystemOriginator();
    var careTaker = new AirConditionSystemCareTaker();

    originator.setState("low");
    var stateLow = originator.saveState(careTaker);
    originator.setState("medium");
    var stateMedium = originator.saveState(careTaker);
    originator.setState("high");
    var stateHigh = originator.saveState(careTaker);

    System.out.printf("""
            Current Air-Condition System state:'%s'%n""",
                originator.getState());

    originator.restoreState(careTaker.getMemento(stateLow));
    System.out.printf("""
            Restored position:'%d', Air-Condition System
                state:'%s'%n""", stateLow,
                    originator.getState());
}
```

Here's the output:

```
Memento Pattern, air-condition system
Current Air-Condition System state:'high'
Restored position:'0', Air-Condition System state:'low'
```

Example 5.12 – Each state is memorized and the driver can restore it on demand

The `AirConditionSystemCareTaker` instance, playing the role of a memento provider, contains links to already used states (*Example 5.13*):

```
final class AirConditionSystemCareTaker {
    private final List<SystemMemento> memory = new
        ArrayList<>();
    ...
    int add(SystemMemento m) {... }

    SystemMemento getMemento(int i) {... }
}
```

Example 5.13 – Each state is remembered with an identifier for restoration

The `AirConditionSystemOriginator` instance considers creating a memento state and restoring the previous one from the memento object. The client is required to remember the provided state identifier to ask the caretaker for a memento state (*Example 5.14*):

```
final class AirConditionSystemOriginator {
    private String state;

    ...

    int saveState(AirConditionSystemCareTaker careTaker){
        return careTaker.add(new SystemMemento(state));
    }

    void restoreState(SystemMemento m){
        state = m.state();
    }
}
```

Example 5.14 – The originator holds the mutable states and updates the caretaker

The program allows the client to only work on a few considered states without creating any other instances (*Figure 5.7*):

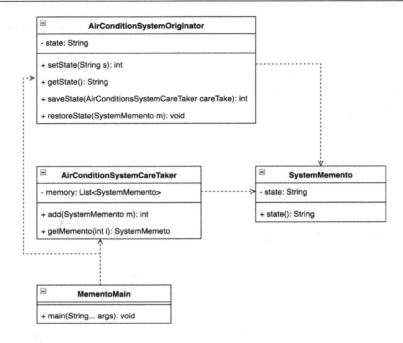

Figure 5.7 – The UML class diagram showing that only a few classes are involved

Conclusion

The memento pattern is very useful when a program needs to perform any undo operation or rewind the timeline. It provides transparent implementation and the separation of logic to enforce the sustainability of the code base.

Let us see the program behaving as expected and examine this in the next section.

Avoiding a null pointer exception state with the null object pattern

The null object pattern provides a way to gracefully deal with an unidentified object and not cause unexpected or undefined program behavior.

Motivation

Instead of using the Java `null` construct to indicate the absence of an object, consider introducing the Null object pattern. A null object is considered to belong to a specific family of objects. The object implements the expected interface, but implementing its methods does not cause any actions.

The advantage of this approach over using an undefined null reference is that the null object is very predictable and has no side effects: it does nothing. It also attempts to eliminate the unpleasant null pointer exception.

Finding it in the JDK

The traditionally mentioned `java.base` module and the `Collection` framework, located in the `java.util` package, define the `Collections` utility class. This class contains an internal private `EmptyIterator` class to serve an `elementless` iterator instance.

Another nice example can be found in the `java.io` module and package. The abstract class, `InputStream`, defines the `nullInpuStream` method that serves the input stream with zero bytes.

Sample code

Let us examine the usage of the null object pattern more closely. Today's vehicles contain a dramatic number of different sensors. In order to take advantage of more than just the functionality of the Stream API, it is quite useful to define a null object that contains the sensor type and that the program can transparently respond to (*Example 5.15*):

```java
public static void main(String[] args) {
    System.out.println("Null Object Pattern, vehicle
        sensor");

    var engineSensor = VehicleSensorsProvider
        .getSenorByType("engine");
    var transmissionSensor = VehicleSensorsProvider
        .getSenorByType("transmission");
    System.out.println("Engine Sensor: " + engineSensor);
    System.out.println("Transmission Sensor: " +
        transmissionSensor);
}
```

Here's the output:

```
Null Object Pattern, vehicle sensor
Engine Sensor: Sensor{type='engine'}
Transmission Sensor: Sensor{type='not_available'}
```

Example 5.15 – The client receives information that the requested sensor is not available as a NullSensor instance

A `VehicleSensorProvider` instance always returns a result of the expected type, and implementing the pattern is very straightforward (*Figure 5.8*):

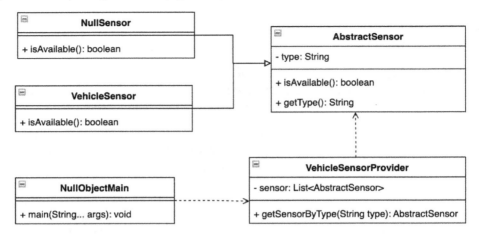

Figure 5.8 – The UML class diagram showing the relationships used in the type maintenance

Conclusion

The examples showed that the pattern can not only improve the maintainability of the code base but also reduce unwanted runtime states, such as a null pointer exception.

An undefined program state can be resolved using an approach that we'll explore in the next section.

Keeping all interested parties informed using the observer pattern

The observer pattern is sometimes known as the producer-customer pattern. Again, this is a very common use case that appears across applications and is therefore mentioned in the GoF's book.

Motivation

A pattern represents a direct relationship between objects. One object has the role of the producer. A producer may have many customers to whom the information should be delivered. These objects are sometimes called receivers. When an observer changes its state, all the registered clients are informed of this change. In other words, any changes that occur to the object will cause the observers to be notified.

Finding it in the JDK

The observer pattern is another fairly commonly used pattern across JDK modules. An example is the `Observer` interface from the `java.base` module and the `java.util` package. Although the interfaces have already been deprecated, they are still used through the `Observable` class in the compiler's implementations.

Sample code

Let us examine temperature control at different locations in the vehicle. The `VehicleSystem` instance should always inform all the interested parties about the temperature goal that each system can adjust itself to (*Example 5.16*):

```
public static void main(String[] args) {
    System.out.println("Observer Pattern, vehicle
        temperature senors");
    var temperatureControlSystem = new VehicleSystem();
    new CockpitObserver(temperatureControlSystem);
    new EngineObserver(temperatureControlSystem);

    temperatureControlSystem.setState("low");
}
```

Here's the output:

```
Observer Pattern, vehicle temperature senors
CockpitObserver, temperature:'11'
EngineObserver, temperature:'4'
```

Example 5.16 – Each subsystem adjusts its temperature according to the global setup

The `SystemObserver` abstract class not only denotes the subsystem under consideration using the construction of sealed classes but also provides a basic template for constructing the intended subsystem (*Example 5.17*):

```
sealed abstract class SystemObserver permits
    CockpitObserver, EngineObserver {
    protected final VehicleSystem system;
    public SystemObserver(VehicleSystem system) {
        this.system = system;
    }
```

```
        abstract void update();
    }
```

Example 5.17 – The newly added subsystem follows the generic template to enforce the maintainability

Each instance contains a reference to the main system that controls the temperature target (*Figure 5.9*):

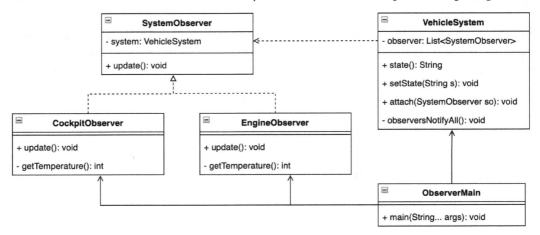

Figure 5.9 – A UML class diagram emphasizes the relationships between systems

Conclusion

The observer pattern is another very powerful one: it allows the client to keep all stakeholders informed without the need to change or understand the implementation. The pattern properly encapsulates and decouples the logic and allows the use of configurable processes at runtime.

The next section shows how to solve linked processes separately.

Dealing with instance stages by using the pipeline pattern

The pipeline pattern can make a significant contribution to improving the organization of multiple downstream operations.

Motivation

This pattern improves data processing in a series of stages by providing an initial input and passing the processed output along for use in subsequent stages. The processing elements are arranged in a continuous pipeline so that the output of one is the input of another, similar to how a physical pipe works. A pipeline pattern can provide some kind of buffering between successive members, represented by object instances. The information that flows through these pipes is often a stream of records.

Finding it in the JDK

The most obvious example of the use of the pipeline pattern is the `Stream` interface and its implementations. The interface is part of the Stream API and is shipped together with the `java.base` module and the `java.util.stream` package.

Sample code

Let's imagine a sequence of processes that need to be carried out in a vehicle, define them, and put them in sequence. We then initialize a `SystemElement` container that collects information about the results of each process (*Example 5.18*):

```
public static void main(String[] args) {
    System.out.println("Pipeline Pattern, vehicle turn on
        states");
    var pipeline = new PipeElement<>(new EngineProcessor())
            .addProcessor(new BreakProcessor())
            .addProcessor(new TransmissionProcessor());

    var systemState = pipeline.process(new
        SystemElement());
    System.out.println(systemState.logSummary());
}
```

Here's the output:

```
Pipeline Pattern, vehicle turn on states
engine-system,break-system,transmission-system
```

Example 5.18 – Each process result is considered in the final summary

The basic construction is `PipeElement`, which defines not only input types but also outputs. Moreover, it defines the order of information processing (*Example 5.19*):

```
class PipeElement<E extends Element, R extends Element> {
    private final Processor<E, R> processor;
```

```
. . .
    <O extends Element> PipeElement<E, O> addProcessor
        (Processor<R, O> p){
        return new PipeElement<>(input -> p.process
            (processor.process(input)));
    }


    R process(E inputElement){
        return processor.process(inputElement);
    }
}
```

Example 5.19 – The addProcessor method defines the order of the processor's process method execution

Each processor implementation can be considered a functional interface construct, and the `Element` implementation can be changed on demand without breaking the pipeline base code (*Figure 5.10*):

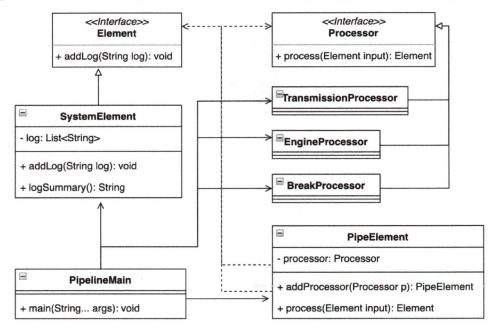

Figure 5.10 – The UML class diagram showing how pipeline type safety is maintained

Conclusion

The presented example shows the advantages of a clear separation of processes contributing to the final result. The pipeline pattern has the potential to create complex operational sequences that can be easily tested and also dynamically changed.

Let's explore how each intended element can change its state in the next section.

Changing object behavior with the state pattern

The behavior state pattern defines the procedure for influencing an object's internal processes based on the mutation of its internal state. This pattern is part of the GoF's book.

Motivation

Object states can be thought of as representing the concept of a finite machine. A pattern allows an object to change its behavior when its internal state changes. The state pattern enforces that an object describes its internal states with specific classes, and maps responses to those states to specific instances.

Finding it in the JDK

Usage of the state pattern can be found in the implementation of the `jlink` plugin, the `jdk.jlink` module, and the `jdk.tools.jlink.plugin` package. The interface plugin defines a nested enum class, `State`, whose values are references to the states in question.

Sample code

The next example considers that each vehicle has different states that are well identified (*Example 5.20*):

```java
public static void main(String[] args) {
    System.out.println("State Pattern, vehicle turn on
        states");
    ...
    var initState = new InitState();
    var startState = new StartState();
    var stopState = new StopState();

    vehicle.setState(initState);
    System.out.println("Vehicle state2:" +
        vehicle.getState());
    vehicle.setState(startState);
```

```
        System.out.println("Vehicle state3:" +
            vehicle.getState());
    vehicle.setState(stopState);
        System.out.println("Vehicle state4:" +
            vehicle.getState());


}
```

Here's the output:

```
State Pattern, vehicle turn on states
Vehicle state2:InitState{vehicle=truck}
Vehicle state3:StartState{vehicle=truck}
Vehicle state4:StopState{vehicle=truck}
```

Example 5.20 – Vehicle states are nicely encapsulated and separated from the client logic

Each Vehicle state considered can be developed independently and properly separated from the client logic (*Figure 5.11*):

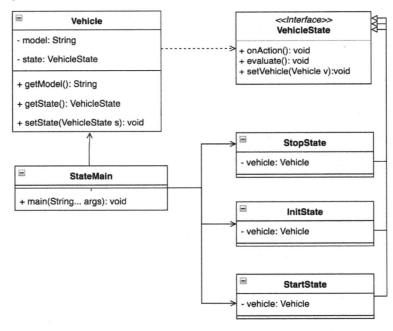

Figure 5.11 – The UML class diagram showing the relation between the states considered

Conclusion

The state pattern shows the advantages of concretely representing a state as a dedicated instance of an object. Not only does it improve testability but it also greatly contributes to the maintainability of the underlying code, as each state is clearly encapsulated and conforms to the single responsibility principle according to the SOLID concept. Program execution becomes transparent to the client without implementing any additional exception-handling logic.

Each state can correspond to a specific program behavior or runtime interaction. Let's dive deeper into this in the next section.

Using the strategy pattern to change object behavior

The strategy pattern can sometimes be called a policy pattern because it establishes precise steps for runtime execution in a particular situation or state. This pattern is a part of the GoF's book.

Motivation

The strategy pattern represents a family of algorithms where each one is properly encapsulated. It defines the interchangeability of algorithms to which a particular object can respond. This strategy allows the algorithm to change independently of the clients using it and allows the client to choose the most appropriate one on the fly. In other words, the code allows the client to attach various strategy objects that affect the behavior of the program.

Finding it in the JDK

The strategy pattern is another pattern commonly used without being aware of its use. The `Collection` framework from the `java.base` module and the `java.util` package implements the `Comparator` class. This class is often used for sorting purposes, such as the implementation of the `Collections.sort()` method.

Another possibly even more widely used implementation is the `map` or `filter` methods introduced by the Stream API, which also comes from the `java.base` module but in the `java.util.stream` package.

Sample code

Suppose a driver has multiple driving licenses required for specific types of vehicles. Each vehicle requires a slightly different driving strategy (*Example 5.21*):

```
public static void main(String[] args) {
    System.out.println("Strategy Pattern, changing
        transport options");
```

```
        var driver = new VehicleDriver(new CarStrategy());
        driver.transport();
        driver.changeStrategy(new BusStrategy());
        driver.transport();
        driver.changeStrategy(new TruckStrategy());
        driver.transport();
    }
```

Here's the output:

```
Strategy Pattern, changing transport options
Car, four persons transport
Bus, whole crew transport
Truck, transporting heavy load
```

Example 5.21 – The VehicleDriver instance can change the transport strategy at the runtime

The VehicleDriver instance only holds the reference to the currently used TransportStrategy instance (*Example 5.22*):

```
class VehicleDriver {
    private TransportStrategy strategy;
    VehicleDriver(TransportStrategy strategy) {
        this.strategy = strategy;
    }
    void changeStrategy(TransportStrategy strategy){
        this.strategy = strategy;
    }
    void transport(){
        strategy.transport();
    }
}
```

Example 5.22 – The VehicleDriver instance communicates with strategy through the visible method

The client can decide which strategy to use at runtime. Each strategy is properly encapsulated (*Figure 5.12*):

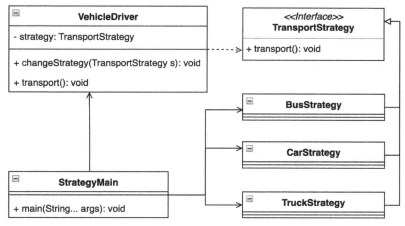

Figure 5.12 – The UML class diagram showing how simply a new strategy can be defined

Conclusion

This trivial example showed a nicely isolated strategic pattern in action. Drivers can change their abilities based on the type of vehicle provided. This pattern's ability to separate its logic from the rest of the code base makes it perfectly suited to implementing complex algorithms or operations that should not be exposed to clients.

Many running events have a general basis. Let's explore how to deal with this kind of situation in the next section.

Standardizing processes with the template pattern

The template method pattern unifies the generalization of intensive actions with a templating approach. The template pattern was recognized early and considered as part of the GoF's book.

Motivation

The template method pattern is based on identifying similarly used steps. These steps define the skeleton of an algorithm. Each operation can defer its steps to specific subclasses. The template method introduces subclasses to redefine certain parts of an algorithm without changing its structure. A template can be used to execute the internal methods in the desired order.

Finding it in the JDK

Java uses input or output byte streams defined by the I/O API located in the `java.base` module and the `java.io` package. The `InputStream` class contains an overloaded `read` method that represents a byte-handling template. It's a similar approach to the `OutputStream` class defining an overloaded `write` method.

Another use of the template pattern can be found in the `Collection` framework, which resides in the same module and the `java.util` package. The abstract `AbstractList`, `AbstractSet`, and `AbstractMap` classes implement the `indexOf` and `lastIndexOf` methods with different templates – for example, `AbstractList` uses `ListIterator`, in comparison to the common `Iterator` interface implementation.

Sample code

Let us examine how the template method pattern can simplify creating a new sensor (*Example 5.23*):

```
public static void main(String[] args) {
    System.out.println("Template method Pattern, changing
        transport options");
    Arrays.asList(new BreaksSensor(), new EngineSensor())
            .forEach(VehicleSensor::activate);


}
```

Here's the output:

```
Template method Pattern, changing transport options
BreaksSensor, initiated
BreaksSensor, measurement started
BreaksSensor, data stored
BreaksSensor, measurement stopped
EngineSensor, initiated
EngineSensor, measurement started
EngineSensor, data stored
EngineSensor, measurement stopped
```

Example 5.23 – The template provides generic activation steps valid for each sensor

The `VehicleSensor` abstract class represents the core element of the example by defining a final `activate` method (*Example 5.24*):

```
abstract sealed class VehicleSensor permits BreaksSensor,
    EngineSensor {
    abstract void init();
    abstract void startMeasure();
    abstract void storeData();
    abstract void stopMeasure();

    final void activate(){
        init();
        startMeasure();
        storeData();
        stopMeasure();
    }
}
```

Example 5.24 – The activate() template method defines the steps for each implementation

In other words, the template method pattern also describes an approach to expanding the vehicle's sensor base (*Figure 5.13*):

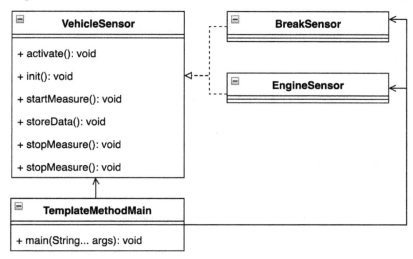

Figure 5.13 – The UML class diagram highlighting the simplicity of adding a new sensor

Conclusion

The template method pattern demonstrates its great advantages for generalized actions. It seamlessly separates the internal logic from the client and provides transparent and generic steps for performing actions. It is easy to maintain the code base or to discover potential issues within it.

The runtime environment can be complex. It's always good to be aware of which instances are present. We will find out how to do this in the next section.

Executing code based on the object type using the visitor pattern

The visitor pattern introduces the separation of algorithm execution from the object instance in question. This pattern is mentioned in the GoF's book.

Motivation

The visitor pattern allows a client to define a new operation without changing the instance of the class it is working on. This pattern provides a way to separate the underlying code from the object structure. The separation practically results in providing the ability to add new operations to an existing object without any modifications to its structure.

Finding it in the JDK

Usage of the visitor pattern can be found in the `java.base` module and the `java.nio.file` package. The `FileVisitor` interface used by the `Files` utility class and its `walkFileTree` method uses a pattern to traverse the directory structure and associated files.

Sample code

A vehicle's security normally relies on the robustness of its sensors. The example shows how to ensure the presence of each specific sensor (*Example 5.25*):

```java
public static void main(String[] args) {
    System.out.println("Visitor Pattern, check vehicle
        parts");
    var vehicleCheck = new VehicleCheck();
    vehicleCheck.accept(new VehicleSystemCheckVisitor());
}
```

Here's the output:

```
Visitor Pattern, check vehicle parts
BreakCheck, ready
BreakCheck, ready, double-check, BreaksCheck@23fc625e
EngineCheck, ready
EngineCheck, ready, double-check, EngineCheck@3f99bd52
SuspensionCheck, ready
SuspensionCheck, ready, double-check,
    SuspensionCheck@4f023edb
VehicleCheck, ready
VehicleCheck, ready, double-check, VehicleCheck@3a71f4dd
```

Example 5.25 – The client also double-checks each sensor's presence

The VehicleSystemCheackVisitor class defines an overloaded implementation of the visit method. Each particular sensor instance can be considered simply by overloading the visit method (*Example 5.26*):

```
class VehicleSystemCheckVisitor implements  CheckVisitor{

    @Override
    public void visit(EngineCheck engineCheck) {
        System.out.println("EngineCheck, ready");
        visitBySwitch(engineCheck);
    }
    private void visitBySwitch(SystemCheck systemCheck){
        switch (systemCheck){
        case EngineCheck e -> System.out.println
            ("EngineCheck, ready, double-check, " + e);
        ...
        default -> System.out.println(
            "VehicleSystemCheckVisitor, not implemented");
        }
    }
    ....
}
```

Example 5.26 – Pattern matching in the instanceof concept can enforce code maintainability

Each system check correctly registers the intended sensor and increases confidence in the vehicle's safety systems (*Figure 5.14*):

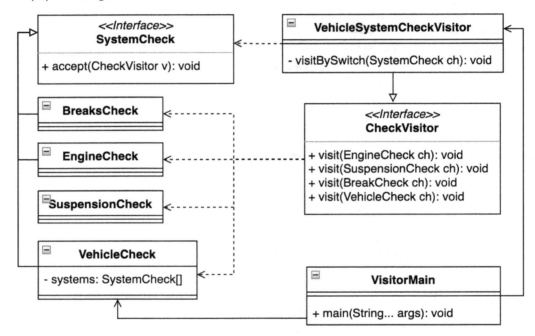

Figure 5.14 – The UML class diagram for the vehicle sensors and their abstractions

Conclusion

This example showed how the `VehicleCheck` system can ensure the presence of each part. Each control is self-contained and new ones can be easily added or removed as needed. The downside is that it requires creating a dedicated instance for each type of control. This also means that at least two classes reference a similar state in the hierarchy. Another advantage or disadvantage is that the pattern does not cause the compilation to fail when a new element that may be discovered at runtime is added. Potential duplication can be overcome and type safety – following the maintenance of the Liskov substitution principle, which is violated by a standard visitor pattern – can be ensured through the utilization of a newly added `switch` statement with pattern matching and a number of other improvements. *Example 5.26* shows the `visitBySwitch` method, which accepts `SystemCheck` objects as input.

Having explored the visitor pattern, we've come to the end of the chapter – let's briefly summarize what we've learned.

Summary

In this chapter, we learned about the importance of the runtime environment and the dynamic nature of program execution. Behavioral design patterns can improve a program's interaction with the internal parts of the Java platform. The JIT compiler can handle dynamic bytecode translation at runtime better, or a garbage collector can perform more efficient memory reclamation.

Most of these design patterns comply with SOLID principles – only the visitor pattern has left some room for thought. However, recently added improvements to the Java platform can help overcome this. Whether sealed classes, `switch` statements, pattern-matching enhancements, or records, the platform provides a solid foundation for strengthening the program's immutability and code stability and simplifying the use of design patterns. Some of them may come out of the box, such as a factory method and `switch-case` statement enhancements.

In this chapter, we learned how to solve challenges at runtime using examples. We explored how to process a chained task and command the required actors. The formula interpreter translates the text into objects and we figured out how to iterate over them. The mediator pattern centralizes complex communication between objects. We learned how to avoid a null-pointer exception and give an undefined object a type using the null object pattern.

The pipeline formula processes a collection of clients sequentially. We explored how to change the state of particular actors and reviewed how to monitor these changes with the observer pattern. The last pattern we learned about was the visitor pattern, which showed us how to perform a specific operation based on the object type.

With the knowledge gained from behavioral patterns, we have added the missing piece to the complete lifecycle of a single-thread program. This includes the creation of objects, the programming structure for working with these objects, and the dynamic behavior and communication between these objects at runtime.

Although the intended program starts from the main thread and may meet be single-threaded as required, neither the Java platform nor most business requirements are single-threaded. The shared nature of tasks lends itself to multi-threaded communication. This can be done using various frameworks. As we will see in the next chapter, we can explore some common concurrency patterns to solve the most common challenges in this regard. Let's shake it up!

Questions

1. What principle is broken by the standard visitor pattern?
2. Which pattern helps us traverse over elements in a collection without knowing the type?
3. Is there a pattern that allows us to change an instance's behavior at runtime?
4. Which pattern helps the runtime transparently identify an undefined state?
5. What are the most used patterns by the Java Stream API?

6. Is there a way to notify all clustered clients at runtime?

7. Which pattern can be used to implement callbacks?

Further reading

- *Design Patterns: Elements of Reusable Object-Oriented Software* by Erich Gamma, Richard Helm, Ralph Johnson, and John Vlissides, Addison-Wesley, 1995

- *Design Principles and Design Patterns* by Robert C. Martin, Object Mentor, 2000

- *JEP-358: Helpful NullPointerExceptions* (https://openjdk.org/jeps/358)

- *JEP-361: Switch Expression* (https://openjdk.org/jeps/361)

- *JEP-394: Pattern Matching for instanceof* (https://openjdk.org/jeps/394)

- *JEP-395: Records* (https://openjdk.org/jeps/395)

- *JEP-405: Sealed Classes* (https://openjdk.org/jeps/405)

- *JEP-409: Sealed Classes* (https://openjdk.org/jeps/409)

Part 3: Other Essential Patterns and Anti-Patterns

This part covers design principles and patterns for building highly concurrent applications. It also discusses several anti-patterns, meaning inappropriate software design solutions to given challenges.

This part contains the following chapters:

- *Chapter 6, Concurrency Design Patterns*
- *Chapter 7, Understanding Common Anti-Patterns*

6

Concurrency Design Patterns

Previous chapters on creational, structural, and behavioral patterns proposed a design that concerns the base code. Their main focus was on the maintainable base code that operates in the main single application thread. In other words, the generated byte code is executed in a defined sequence to achieve the desired results.

Nowadays, business requirements have shifted the application expectations described by the GoF's book over the years more and more within a concurrent and parallel world. This has been succeeded by a massive improvement in hardware.

The Java platform provides concurrency functionality under the hood from the very beginning. The Flight Recorder tool of Mission Control set helps collect data points about thread behavior and displays them visually, improving our awareness of application dynamics. In this chapter, we are going to examine some of the most common scenarios in the information technology industry:

- Decoupling a method execution with an active object pattern
- Non-blocking tasks using an async method invocation pattern
- Delaying execution until the previous task has been completed with the balking pattern
- Providing a unique object instance with a double-checked locking pattern
- Using purposeful thread blocking via a read-write lock pattern
- Decoupling the execution logic with the producer-consumer pattern
- Executing isolated tasks with the scheduler pattern
- Effective thread utilization with the thread-pool pattern

By the end of this chapter, we will have built a solid foundation for understanding the concurrency possibilities of the Java platform and starting to apply them effectively.

Technical requirements

You can find the code files for this chapter on GitHub at `https://github.com/PacktPublishing/Practical-Design-Patterns-for-Java-Developers/tree/main/Chapter06`.

Decoupling a method execution with an active object pattern

The active object design pattern separates and defers method execution from method invocation by running its own control thread.

Motivation

The active object pattern introduces a transparently concurrent model to the application. It creates and starts an internal thread that executes the required logical, critical section. An active object instance exposes a public interface that a client can use to run an encapsulated critical section. An external, client-initiated event is queued and ready to execute. The execution step is performed by the internal scheduler. The result can be passed to the appropriate handler in a callback style.

Sample code

Let us introduce an example of a moving vehicle with a radio system (*Example 6.1*):

```
public static void main(String[] args) throws Exception {
    System.out.println("Active Object Pattern, moving
        vehicle");
    var sportVehicle = new SportVehicle("super_sport");
    sportVehicle.move();
    sportVehicle.turnOnRadio();
    sportVehicle.turnOffRadio();
    sportVehicle.turnOnRadio();
    sportVehicle.stopVehicle();
    sportVehicle.turnOffRadio();
    TimeUnit.MILLISECONDS.sleep(400);
    System.out.println("ActiveObjectMain, sportVehicle
    moving:" + sportVehicle.isMoving());
}
```

Here's the output:

```
Active Object Pattern, moving vehicle
MovingVehicle:'super_sport', moving
MovingVehicle:'super_sport', radio on
MovingVehicle:'super_sport', moving
MovingVehicle:'super_sport', stopping, commands_active:'3'
MovingVehicle:'super_sport', stopped
ActiveObjectMain, sportVehicle moving:false
```

Example 6.1 – The SportVehicle instance allows the client to create an event by using its public methods

The newly created abstract class, MovingVehicle, defines public methods – move, turnOnRadio, turnOffRadio, and stopVehicle. In addition to the control thread, the class defines a conditional queue for incoming events (*Example 6.2*):

```
abstract class MovingVehicle {
    private static final AtomicInteger COUNTER = new
        AtomicInteger();
    private final BlockingDeque<Runnable> commands;
    private final String type;
    private final Thread thread;
    private boolean active;

    MovingVehicle(String type) {
        this.commands = new LinkedBlockingDeque<>();
        this.type = type;
        this.thread = createMovementThread();
    }
    ...
  private Thread createMovementThread() {
    var thread = new Thread(() -> {
        while (active) {
            try {
                var command = commands.take();
                command.run();

                . . .
```

```
        }
    });
    thread.setDaemon(true);
    thread.setName("moving-vehicle-" +
        COUNTER.getAndIncrement());
    return thread;
    ...
}
```

Example 6.2 – MovingVehicle contains an active flag for purpose of scheduling events

Events in the queue are received and fired based on an internal period. `LinkedBlockingDeque` provides additional functions for inserting or removing elements from the top or bottom, which is useful when the vehicle needs to be stopped. The `StopVehicle` event has priority over the radios (*Example 6.3*):

```
abstract class MovingVehicle {
    ...
    void turnOffRadio() throws InterruptedException {
        commands.putLast(() -> {...});
    }
    void stopVehicle() throws InterruptedException {
        commands.putFirst(() -> {...});
    }
    ...
}
```

Example 6.3 – The received events are added to the queue conditionally

A lifecycle of the `SportVehicle` instance does not interfere with the main application thread. It is predictable and does not block the application (*Figure 6.1*):

moving-vehicle-0

18/11/2022, 19:15:25.300.000

Figure 6.1 – The moving-vehicle thread shows the sequence of commands

The components introduced in the example seamlessly cooperate (*Figure 6.2*):

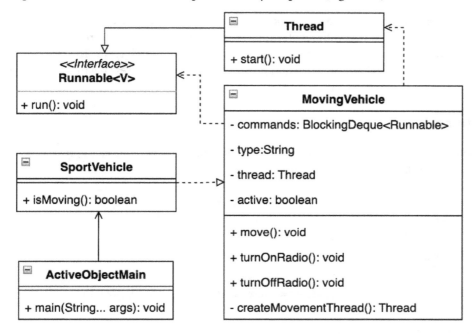

Figure 6.2 – The UML class diagram shows the SportVehicle class's relation to the Java concurrency features

Conclusion

A well-developed active object pattern respects the SOLID design approaches because it encapsulates the critical parts and only exposes the control interface required. The instance does not interfere with the application and the whole approach can be generalized to the desired level. An active object can be a good candidate for introducing a concurrency model into an application, but there are a few challenges to keep in mind. One of these challenges is the number of possible application threads, where a high number can make the application fragile or lead to instability as it depends on available resources.

Let's explore the asynchronous nature of events in the next section.

Non-blocking tasks using async method invocation pattern

The asynchronous method invocation pattern is a way to solve the challenge of not penalizing the main process thread with possibly time-consuming tasks.

Motivation

Asynchronous method invocation patterns introduce the ability to receive a result by a callback from an asynchronously running task without blocking the main process thread. The pattern presents the threading model and level of parallelism for processing the required task types. The task results are processed by dedicated callback handlers and provided to the main process regardless of the task's execution time. These handlers may already belong to the main process.

Sample code

Let us look at a trivial scenario of several vehicle temperature sensors required to provide results to the driver, which is the client (*Example 6.4*):

```
public static void main(String[] args) throws Exception {
    System.out.println("Async method invocation Pattern,
        moving vehicle");
    var sensorTaskExecutor = new
        TempSensorExecutor<Integer>();
    var tempSensorCallback = new TempSensorCallback();

    var tasksNumber = 5;
    var measurements = new ArrayList<SensorResult
        <Integer>>();
    System.out.printf("""
                AsyncMethodMain, tasksNumber:'%d' %n""",
                    tasksNumber);
    for(int i=0; i<tasksNumber; i++) {
        var sensorResult = sensorTaskExecutor.measure(new
            TempSensorTask(), tempSensorCallback);
        measurements.add(sensorResult);
    }
    sensorTaskExecutor.start();
    AsyncMethodUtils.delayMills(10);
    for(int i=0; i< tasksNumber; i++){
        var temp = sensorTaskExecutor.stopMeasurement
            (measurements.get(i));
        System.out.printf("""
                AsyncMethodMain, sensor:'%d'
```

```
                    temp:'%d'%n""", i, temp);
    }
```

Here's the output:

```
Async method invocation Pattern, moving vehicle
AsyncMethodMain, tasksNumber:'5'
SensorTaskExecutor, started:5
...
TempSensorTask,n:'4' temp:'5', thread:'thread-3'
TempSensorTask,n:'3' temp:'26', thread:'thread-0'
TemperatureSensorCallback, recorded value:'26',
    thread:'main'
AsyncMethodMain, sensor:'0' temp:'26'
...
```

Example 6.4 – The example task temp:26, is asynchronously executed in thread-0 thread

The instance of `TempSensorCallback` that monitors all results resides in the `main` process thread (*Figure 6.3*):

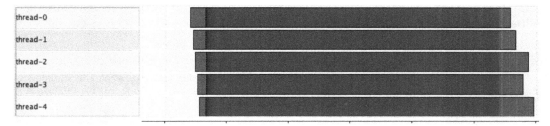

Figure 6.3 - TemperatoreSensorCallback instance is called
asynchronously, therefore different thread finish times

`TempSensorTask` instances are handled by custom `TempSensorExecutor` instances, which not only provides control over initiated threads, but can also terminate long-running measurements of a particular sensor by providing a task reference. The `TempSensorExecutor` instance exposes a measure public method that provides a `TempSensorResult` instance of a long-running task (*Example 6.5*):

```
class TempSensorExecutor<T> implements SensorExecutor<T> {
    ...
```

```
@Override
public SensorResult<T> measure(Callable<T> sensor,
    SensorCallback<T> callback) {
    var result = new TempSensorResult<T>(callback);
    Runnable runnable = () -> {
        try {
            result.setResult(sensor.call());
        } catch (Exception e) {
            result.addException(e);
        }
    };
    var thread = new Thread(runnable, "thread-" +
        COUNTER.getAndIncrement());
    thread.setDaemon(true);
    threads.add(thread);
    return result;
    }
}
```

Example 6.5 – Each new thread specific long-term measurement will pass its result to the callback handler

The nature of processing the information served by multiple temperature sensors is clearly parallel. The asynchronous method invocation patterns provide a very small set of classes to solve this challenge (*Figure 6.4*):

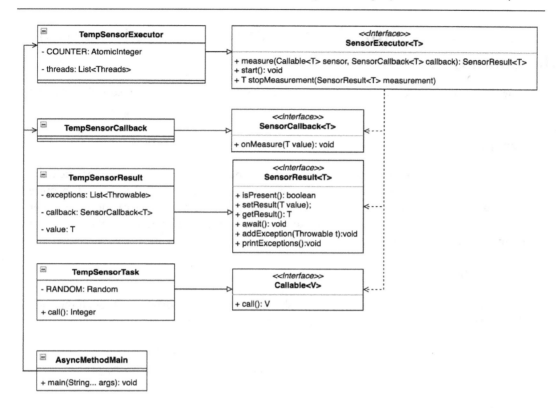

Figure 6.4 – The UML class diagram draws how data are acquired from temperature sensors

Conclusion

The given examples clearly show how to handle a long-running task with a preliminary detachment from the main processing thread. In other words, it is not causing by delas. The Java platform provides multiple options to create this pattern. One of them is to employ the `Callable` interface and send an instance to the `ExecutorService` using its `submit` method. The `submit` method returns a result that implements the `Future` interface. The `Future` has similarities to the sample `TempSensorResult` instance but does not provide a callback function that needs to be handled differently. Another possibility can be explored using `CompletableFuture`, which not only exposes the `supplyAsync` method, but also provides many other useful functions. All of the suggestions mentioned can be found in the `java.base` module and the `java.util.concurrent` package.

The next section shows how to delay the execution of a task until the previous one is complete; let's get to it.

Delay execution until the previous task is completed with the balking pattern

Sometimes it is required to consider the task state changes to properly execute next task and fulfill the goal.

Motivation

Although the instance mutability is not a desirable state, especially not in the concurrency field the ability to depend on the object state may come handy. The case where multiple threads try to acquire an object to execute its critical sections can be limited by the object state. The state can decide whether the processing time will be used or not in order to coordinate the resources available. For example, a vehicle cannot stop without being in motion.

Sample code

Consider the example of one `Vehicle` instance sharing two groups of drivers. Although there are multiple groups, only one vehicle can operate at a time (*Example 6.6*):

```java
public static void main(String[] args) throws Exception {
    System.out.println("Balking pattern, vehicle move");

    var vehicle = new Vehicle();
    var numberOfDrivers = 5;
    var executors = Executors.newFixedThreadPool(2);
    for (int i = 0; i < numberOfDrivers; i++) {
        executors.submit(vehicle::drive);
    }
    TimeUnit.MILLISECONDS.sleep(1000);
    executors.shutdown();
    System.out.println("Done");
}
```

Here's the output:

```
Balking pattern, vehicle move
Vehicle state:'MOVING', moving, mills:'75',
  thread='Thread[pool-1-thread-2,5,main]'
```

```
Vehicle state:'STOPPED' stopped, mills:'75',
  thread='Thread[pool-1-thread-2,5,main]'
Vehicle state:'MOVING', moving, mills:'98',
  thread='Thread[pool-1-thread-1,5,main]'
...
```

Example 6.6 – Driver groups are represented by provided threads created by ExecutorService

The balking pattern provides a solution in which the critical section of the task is executed based on the Vehicle instance state represented by the VehicleState enum (*Example 6.7*):

```
class Vehicle {
    synchronized void driveWithMills(int mills) throws
        InterruptedException {
        var internalState = getState();
        switch (internalState) {
            case MOVING -> System.out.printf("""
                    Vehicle state:'%s', vehicle in move,
                        millis:'%d', thread='%s'%n""",
                            state, mills, Thread
                                .currentThread());
            case STOPPED -> startEngineAndMove(mills);
            case null -> init();
        }
    }
}
...
```

Example 6.7 – The usage of the synchronized keyword makes the driver groups aware of whether the Vehicle instance is ready to use or not

The driver group threads are blocked and only one is active at a time (*Figure 6.5*):

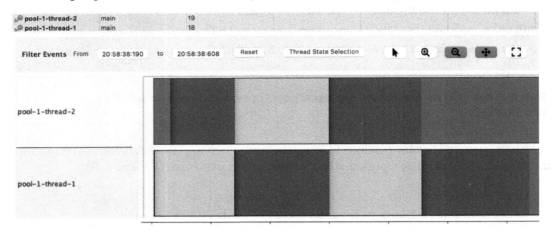

Figure 6.5 – The blue and green colors represent group activity while the other is blocked

The example presented requires a very minimal number of created classes, which are clearly encapsulated (*Figure 6.6*):

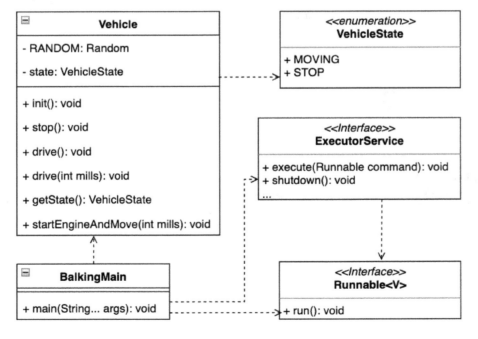

Figure 6.6 – The UML class diagram shows the two most required custom classes, Vehicle and VehicleState

Conclusion

The balking pattern is easy to implement in the Java platform. It is important to keep the Java memory model in mind in order to properly handle object state volatility. It may be particularly useful to consider using atomic types (`AtomicInteger` and `AtomicBoolean`, for example), which automatically come with a happens-before guarantee. This guarantee is part of the Java memory model to maintain memory consistency across the interacting threads, as we learned in *Chapter 2, Discovering the Java Platform for Design Patterns*. Another option to consider is the `volatile` keyword, which comes with a guarantee of equal-value visibility across threads.

The next section examines guaranteed instance uniqueness – let's roll.

Providing a unique object instance with a double-checked locking pattern

The double-checked locking pattern solves the problem of an application requiring only one instance of a particular class at runtime.

Motivation

The Java platform is multi-threaded by default, as we learned in *Chapter 2, Discovering the Java Platform for Design Patterns*. It's not just the garbage collection threads that take care of the main program lifecycle. Different frameworks introduce additional tread models, which may have an unintended impact on a class institution's process. A double-checked locking pattern ensures that only one instance of a class is present at runtime. This state can become challenging in a multi-threaded environment, as it may depend on its implementation.

Sample code

Let's use a simple `Vehicle` instance to demonstrate the importance of a double-checked locking pattern in a multithreading environment. The example presents two different implementations of the singleton pattern. `VehicleSingleton` is expected to keep its promise due to multiple threads accessing the `getInstance` method (*Example 6.8*):

```java
public static void main(String[] args) {
    System.out.println("Double checked locking pattern,
        only one vehicle");
    var amount = 5;
    ExecutorService executor = Executors
        .newFixedThreadPool(amount);
    System.out.println("Number of executors:" + amount);
```

```
        for (int i = 0; i < amount; i++) {
            executor.submit(VehicleSingleton::getInstance);
            executor.submit(VehicleSingletonChecked
                ::getInstance);
        }
        executor.shutdown();
    }
```

Here's the output:

```
Double checked locking pattern, only one vehicle
Number of executors:5
VehicleSingleton, constructor thread:'pool-1-thread-1'
hashCode:'1460252997'
VehicleSingleton, constructor thread:'pool-1-thread-5'
hashCode:'1421065097'
VehicleSingleton, constructor thread:'pool-1-thread-3'
hashCode:'1636104814'
VehicleSingletonChecked, constructor thread:'pool-1-thread-
2' hashCode:'523532675'
```

Example 6.8 – The VehicleSingleton constructor has been called multiple times, which violates the given promise through multiple instantiations (see the hashCode values)

The `ExecutorService` instance provided by `Executors.newFixedThreadPool` receives multiple instances of the `Runnable` interface. The `Runnable` method's implementation represents the critical section of the `getInstance` method's call in both cases (*Figure 5.5*):

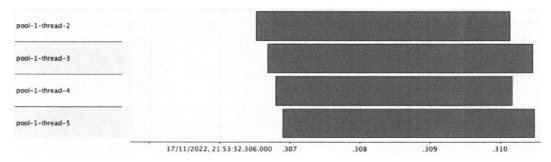

Figure 6.7 – All pool threads continuously execute the getInstance
method and VehicleSingletonCheck is created only once

Both implementations differ in the very small details of the getInstance method implementation (*Example 6.9*):

```
public static VehicleSingleton getInstance(){
    if (INSTANCE == null){
        INSTANCE = new VehicleSingleton();
    }
    return INSTANCE;
}
...
static VehicleSingletonChecked getInstance() {
    if (INSTANCE == null) {
        synchronized (VehicleSingletonChecked.class) {
            if (INSTANCE == null) {
                INSTANCE = new VehicleSingletonChecked();
            }
        }
    }
    return INSTANCE;
}
```

Example 6.9 – The getInstance method's implementation of VehicleSingletonChecked uses a synchronized keyword to ensure the thread stack frame state

In both cases, the UML diagram remains the same (*Figure 6.8*):

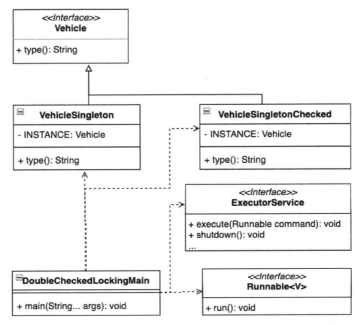

Figure 6.8 – The UML class diagram does not highlight the implementation
details of the double-checked singleton pattern

Conclusion

This example has shown a possible way to implement a double-check locking pattern. The Java platform may also enforce double-checked locking patterns by using an Enum construct, which provides only one element – its INSTANCE object of desired type.

The next section demonstrates how to deal with locking exclusivity.

Using purposeful thread blocking via a read-write lock pattern

A concurrent application may consider granting exclusive access to a critical section just to update the information of the specific instance. This particular challenge can be solved by using a read-write lock pattern.

Motivation

The read-write locking pattern introduces natural exclusivity for lock acquisition. This context is used to differentiate the whether the critical section can be executed. In other words, the write action takes precedence by its nature before reading, as the goal of any reader is to get the most accurate and up-to-date value possible. Under the hood, this means that all readers are blocked when the writer thread is modifying data and unblocked when the writer completes its task.

Sample code

Suppose that multiple sensors inside a vehicle require accurate information about the temperature value, but there is only one temperature device capable of updating the temperature value (*Example 6.10*):

```
public static void main(String[] args) throws Exception {
    System.out.println("Read-Write Lock pattern, writing
        and reading sensor values");
    ReentrantReadWriteLock readWriteLock = new
        ReentrantReadWriteLock();
    var sensor = new Sensor(readWriteLock.readLock(),
        readWriteLock.writeLock());
    var sensorWriter = new SensorWriter("writer-1",
        sensor);
    var writerThread = getWriterThread(sensorWriter);

    ExecutorService executor = Executors.newFixedThreadPool
        (NUMBER_READERS);
    var readers = IntStream.range(0, NUMBER_READERS)
            .boxed().map(i -> new SensorReader("reader-"
                + i, sensor, CYCLES_READER)).toList();
    readers.forEach(executor::submit);
    writerThread.start();
    executor.shutdown();
}
```

Here's the output:

```
Read-Write Lock pattern, writing and reading sensor values
SensorReader read, type:'reader-2', value:'50,
thread:'pool-1-thread-3'
```

```
SensorReader read, type:'reader-0', value:'50,
thread:'pool-1-thread-1'
SensorReader read, type:'reader-1', value:'50,
thread:'pool-1-thread-2'
SensorWriter write, type:'writer-1', value:'26',
thread:'pool-2-writer-1'
SensorReader read, type:'reader-2', value:'26,
thread:'pool-1-thread-3'
...
```

Example 6.10 – The SensorWriter instance that runs its own thread obtains exclusive access to the Sensor instance

Readers continuously read the sensor value without being blocked. The situation changes when the writer enters the game – at which point, readers are blocked and have to wait for the SensorWriter instance to finish (*Figure 6.9*):

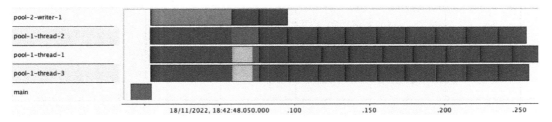

Figure 6.9 – Thread activity highlighting writer lock exclusivity when reader threads are being blocked

The critical section is served by two methods, writeValue and readValue. Both belong to the Sensor class (*Example 6.11*):

```
class Sensor {
    ...
    int getValue() {
        readLock.lock();
        int result;
        try {  result = value; ... } finally {  readLock.
          unlock(); }
        return result;
    }
```

```
void writeValue(int v) {
    writeLock.lock();
    try { this.value = v; ...} finally {
        writeLock.unlock();}
    }
}
```

Example 6.11 – readLock is paused when writeLock is acquired

It is important to note that lock instances reside in the main thread of execution and are acquired by the threads provided by the `ExecutorService` instance (*Figure 6.10*):

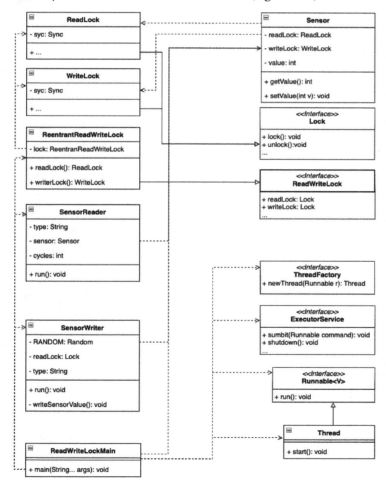

Figure 6.10 – The UML class diagram for a read-write lock pattern

Conclusion

The read-write lock is very powerful and can contribute very positively to the stability of the application. It clearly separates the participant from the critical section code that drives the logic. Each example class can be generalized or adapted according to SOLID design principles upon request.

The JDK defines another approach worth considering for exchanging the sensor value. The `java.base` module package, `java.util.concurrent`, contains the `Exchanger` class, which provides the required synchronization guarantees.

Let's examine another common pattern where the instance is broadcasted to the target.

Decoupling the execution logic with a producer-consumer pattern

The common industrial scenario represents producing and consuming values without blocking the main application thread. The producer-consumer pattern helps to solve this challenge by decoupling the logic and separating the lines.

Motivation

A common industrial scenario involves producing and consuming values without blocking the main execution thread. The producer-consumer pattern is designed exactly to rise to the challenge by decoupling the logic and separating the target receivers.

Sample code

Another scenario is where the vehicle produces multiple events from multiple sources and these events need to be broadcasted and delivered to consumers (*Example 6.12*):

```
public static void main(String[] args) throws Exception{
    System.out.println("Producer-Consumer pattern,
        decoupling receivers and emitters");
    var producersNumber = 12;
    var consumersNumber = 10;
    var container = new EventsContainer(3);

    ExecutorService producerExecutor =
        Executors.newFixedThreadPool(4, new
            ProdConThreadFactory("prod"));
    ExecutorService consumersExecutor = Executors.
```

```
        newFixedThreadPool(2, new ProdConThreadFactory
            ("con"));
    IntStream.range(0, producersNumber)
            .boxed().map(i -> new EventProducer(container))
            .forEach(producerExecutor::submit);
    IntStream.range(0, consumersNumber)
            .boxed().map(i -> new EventConsumer(i,container))
            .forEach(consumersExecutor::submit);
    TimeUnit.MILLISECONDS.sleep(200);
    producerExecutor.shutdown();
    consumersExecutor.shutdown();
}
```

Here's the output:

```
Producer-Consumer pattern, decoupling mess
VehicleSecurityConsumer,event:'Event[number=0, source=pool-
prod-0]', number:'0', thread:'pool-con-0'
VehicleSecurityConsumer,event:'Event[number=1, source=pool-
prod-3]', number:'1', thread:'pool-con-1'
VehicleSecurityConsumer,event:'Event[number=3, source=pool-
prod-1]', number:'2', thread:'pool-con-0'
VehicleSecurityConsumer,event:'Event[number=2, source=pool-
prod-2]', number:'3', thread:'pool-con-1'
...
```

Example 6.12 – Compared to producers, consumers are in the minority, not only in terms of quantity but also in terms of available resources

Each of the `ExecutorService` instances uses a `ProdConThreadFactory` object type to provide meaningful thread names (*Figure 6.11*):

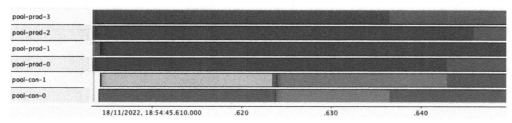

Figure 6.11 – Consumers are in the minority and sometimes may be blocked as the event storage is full

The participant classes are decoupled and ready for extension (*Figure 6.12*):

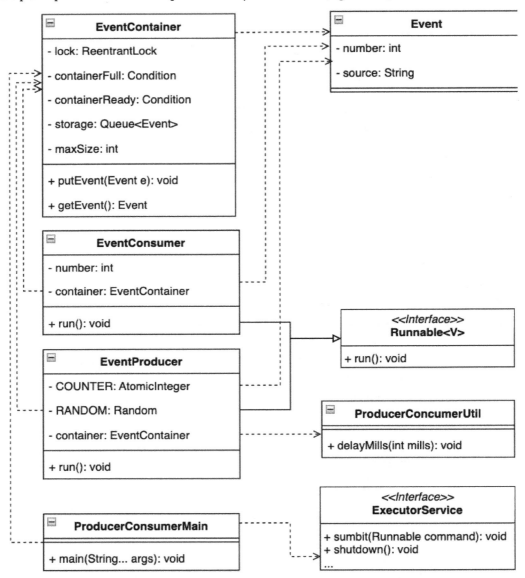

Figure 6.12 – The UML class diagram shows how event classes
are related to the internals of the Java platform

Conclusion

In the field of distributed systems, the production-consumer approach is widely used. It is advantageous for clearly separating and defining groups of event senders and receivers. The groups can be placed in different threads according to the desired thread model.

The JDK 19 release comes with the newly added concept of virtual threads. Virtual threads attempt to simplify the use of core platform threads by introducing thread-like frames, and wrappers. The virtual thread wrappers are scheduled by the JVM to run on available platform threads by using newly added executors – for example, `Executors.newVirtualThreadPerTaskExecutor`. This approach fulfills the definition of the producer-consumer pattern, in which the producer is an application that uses new virtual thread executors and the platform consumes scheduled virtual threads.

Let's uncover the scheduler approach in the next section in more detail.

Executing isolated tasks with the scheduler pattern

An application behaving deterministically can play a critical role in its success. A scheduler pattern can help to achieve the desired goal.

Motivation

Although schedulers are sometimes poorly designed to keep the application busy, their main purpose is important. The importance of using patterns comes to light more with microservices or distributed approaches in which the system is required to behave predictably. The general goal is to determine when a specific task is performed so that the underlying resources are properly used or a budget estimate can be created for the required resources described in site reliability engineering.

Sample code

The following example brings us to temperature measurement. Every vehicle contains temperature sensors in a mechanical or digital form. Temperature sensors play a key role in vehicle operation (*Example 6.13*):

```
public static void main (String [] args) throws Exception {
    System.out.println("Scheduler pattern, temperature
        measurement");
    var scheduler = new CustomScheduler(100);
    scheduler.run();
    for (int i=0; i < 15; i++){
        scheduler.addTask(new SensorTask(
            "temperature-"+i));
```

```
        }

        TimeUnit.SECONDS.sleep(1);
        scheduler.stop();
    }
```

Here's the output:

```
Scheduler pattern, providing sensor values
SensorTask, type:'temperature-0'
,activeTime:'58',thread:'scheduler-1'
SensorTask, type:'temperature-1',
activeTime:'65',thread:'scheduler-1'
SensorTask, type:'temperature-2',
activeTime:'75',thread:'scheduler-1'
...
CustomScheduler, stopped
```

Example 6.13 – The CustomScheduler instance executes a SensorTask instance from the blocking queue every 100 milliseconds

`CustomerScheduler` shows a trivial implementation that administers the execution process (*Figure 6.14*):

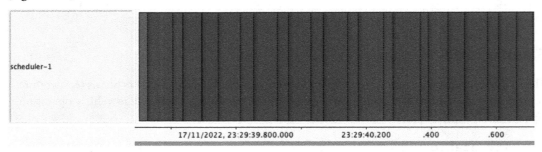

Figure 6.13 – Each task execution has a 100 ms time window allocated

The scheduler instantiation prepares a thread with an active flag to control the lifecycle (*Example 6.14*):

```
CustomScheduler { ...
    CustomScheduler(int intervalMillis) {
    this.intervalMills = intervalMillis;
    this.queue = new ArrayBlockingQueue<>(10);
```

```
    this.thread = new Thread(() -> {
        while (active){
            try {
                var runnable = queue.poll(intervalMillis,
                    TimeUnit.MILLISECONDS);
                ...
                var delay = intervalMillis - runnable
                    .activeTime();
                TimeUnit.MILLISECONDS.sleep(delay);
            } catch (InterruptedException e) {  throw new
                RuntimeException(e); }
        }
        System.out.println("CustomScheduler, stopped");
    }, "scheduler-1");
}
...
```

Example 6.14 – CustomScheduler ensures that the time window is maintained

The task of creating a simple scheduler is trivial, but beyond that, it's good to keep in mind the threading model – as in, where and how execution takes place (*Figure 6.14*):

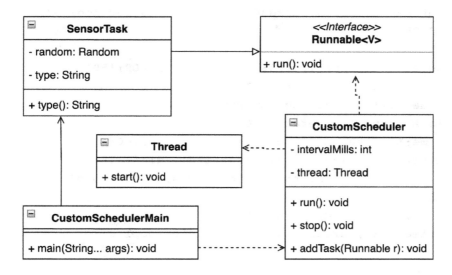

Figure 6.14 – The UML class diagram highlights the CustomScheduler threading model

In the case of the scheduler pattern, it is only fair to mention the second example. The second example uses the built-in JDK functions and their customizations. The planning process is fully managed by the platform. The application example is, again, similar to the first example, temperature measurement (*Example 6.15*):

```java
public static void main(String[] args) throws Exception {
    System.out.println("Pooled scheduler pattern ,
        providing sensor values");
    var pool = new CustomScheduledThreadPoolExecutor(2);

    for(int i=0; i < 4; i++){
        pool.scheduleAtFixedRate(new SensorTask
            ("temperature-"+i), 0, 50,
                TimeUnit.MILLISECONDS);
    }
    TimeUnit.MILLISECONDS.sleep(200);
    pool.shutdown();
}
```

Here's the output:

```
Pooled scheduler pattern, providing sensor values
POOL: scheduled task:'468121027', every MILLS:'50'
POOL, before execution, thread:'custom-scheduler-pool-0',
task:'852255136'
...
POOL: scheduled task:'1044036744', every MILLS:'50'
SensorTask, type:'temperature-1',
activeTime:'61',thread:'custom-scheduler-pool-1'
SensorTask, type:'temperature-0',
activeTime:'50',thread:'custom-scheduler-pool-0'
POOL, after execution, task:'852255136', diff:'56'
POOL, before execution, thread:'custom-scheduler-pool-0',
task:'1342170835'
SensorTask, type:'temperature-2'
,activeTime:'71',thread:'custom-scheduler-pool-0'
...
POOL is going down
```

Example 6.15 – The period is set to 100 ms and the SensorTask instance is reused for each iteration

An extended `CustomScheduledThreadPoolExecutor` instance can provide additional information based on task execution by overriding an available method such as `beforeExecute` or `afterExecute`. Using the JDK internals makes it easy to scale a `SensorTask` instance across threads (*Figure 6.15*):

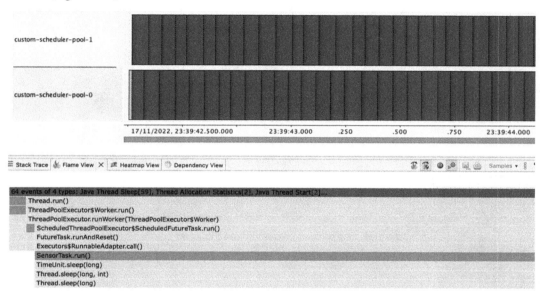

Figure 6.15 – A CustomScheduledThreadPoolExecutor instance facilitates easier
thread management and easier management of other JDK internals

Leveraging the JDK internals for scheduling does not require you to create a customized solution while gaining better visibility into the scheduling cycle (*Figure 6.16*):

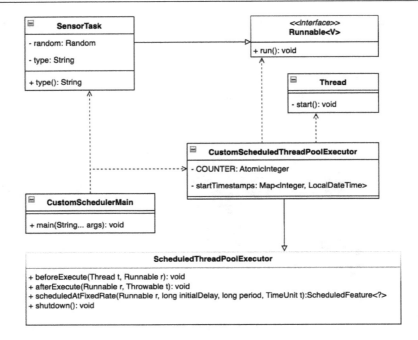

Figure 6.16 – The UML class diagram shows the minimal effort required
to create a custom scheduler with all the internals

Conclusion

Both preset examples show possible uses of the scheduler pattern. Using the JDK internals has a number of advantages to consider. It gives the platform the ability to more efficiently use and optimize available resources, such as the dynamic JIT translation we learned about in *Chapter 2, Discovering the Java Platform for Design Patterns.*

Effective thread utilization using a thread-pool pattern

It is not always necessary to create a new thread for each task, as this can lead to improper resource usage. A thread-pool pattern may be a good solution to this challenge.

Motivation

A short-lived task does not need to create a new thread every time it runs, because each instantiation of a thread is related to the allocation of underlying resources. Wasting resources could result in an application throughput or performance penalty. A better option is described by the thread-pool pattern, which defines the required number of reusable threads to execute a critical section. Specific workers can transparently operate above the critical section code that needs to be executed.

Sample code

Let us imagine again a temperature measurement by sensors with different measurement dynamics (*Example 6.16*):

```java
public static void main (String[] args) throws Exception{
    System.out.println("Thread-Pool pattern, providing
        sensor values");
    var executor = Executors.newFixedThreadPool(5);

    for (int i=0; i < 15; i++){
        var task = new TemperatureTask("temp" + i);
        var worker  = new SensorWorker(task);
        executor.submit(worker);
    }
    executor.shutdown();
}
```

Here's the output:

```
Thread-Pool pattern, providing sensor values
TemperatureTask, type:'temp3', temp:'0', thread:'pool-1-
thread-4'
TemperatureTask, type:'temp4', temp:'7', thread:'pool-1-
thread-5'
TemperatureTask, type:'temp2', temp:'15', thread:'pool-1-
thread-3'
TemperatureTask, type:'temp1', temp:'20', thread:'pool-1-
thread-2'
..
```

Example 6.16 – The thread pool runs the worker temperature measurement task on demand

A thread pool helps to use and manage created threads so that there is always a task to process. This positively affects the application behavior and facilitates planning based on the resources available (*Figure 6.17*):

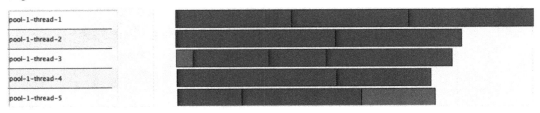

Figure 6.17 – The behavior of the thread pool shows the usage of the created thread

The core example element is `SensorWorker`. The worker implements a `Runnable` interface and is responsible for the `TemperatureTask` evaluation (*Example 6.17*):

```
class SensorWorker implements Runnable {

    ...
    @Override
    public void run () {
        try {task.measure();} catch (InterruptedException
            e) {...}
    }
}
```

Example 6.17 – The SensorTask instance can provide additional logic for the task evaluation wrapped around it

The example implementation does not require any additional custom class types to introduce concurrency (*Figure 6.16*):

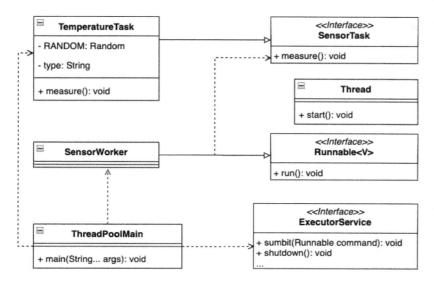

Figure 6.18 – The UML class diagram highlights that all the required
thread pool elements are provided by the Java platform

Conclusion

A thread-pool pattern can provide another acceptable way to introduce concurrency into an application. It not only supports the execution of class types that inherit the Runnable interface but also the Callable interface. Using a Callable interface allows you to create a result through a Future interface. The result of executing a Callable instance into an instance of the Future type is that the execution of the critical section is done asynchronously by the controlling thread. In other words, the time required to produce a result is not known.

The thread-pool pattern is also another SOLID approach to properly structuring your code base to ensure maintainability and resource utilization.

Let us briefly summarize the lessons learned in this chapter.

Summary

This chapter demonstrated some of the most commonly used approaches to solving concurrent problems. It also showed the importance of previously acquired knowledge, concurrent application development requires more precision and discipline to achieve the desired result, similar to the knowledge of Java platform internals discussed in *Chapter 2, Discovering the Java Platform for Design Patterns*.

Each currently adopted pattern forces the creation of a sustainable, clean application code base. Many of them clearly follow and use discussed development approaches such as APIE or SOLID.

The evolution of the Java platform tends to simplify how to approach the platform's concurrency capabilities. One good example has been already mentioned in some of the sections in this chapter. Features such as `CompletableFuture` or `Executors` utils have been around for a while, but upcoming ones might be worth considering. The tentative goal with virtual threads is to increase application throughput while making proper use of the underlying resources and still maintaining threading conveniences such as debugging and providing relevant stack frames. Structure concurrency, on the other hand, attempts to provide a framework for simply designing callbacks while using an imperative code style. In addition to upcoming features that try to improve application throughput or simplify the usage of the concurrency framework, we should not forget about the immutability of the instances served by the `record` type. The `record` type provides a strong state contract due to equality. Instance immutability can play a critical role in thread interactions.

The entire application development can sometimes deviate from the desired goal. Some common symptoms have already been identified in this kind of situation. These signals could signal our attention to reconsider the direction of development.

We will touch on some of them in the next chapter.

Questions

1. What challenges are solved by the double-check singleton pattern?
2. What is the best way to create the desired thread pool with the JDK?
3. Which concurrent design pattern reflects the variability of an instance to process the next step?
4. What is the best pattern for handling a repeatable task?
5. What pattern helps separate the dispatch logic and event handling?

Further reading

- *Design Patterns: Elements of Reusable Object-Oriented Software* by Erich Gamma, Richard Helm, Ralph Johnson, and John Vlissides, Addison-Wesley, 1995

- *Design Principles and Design Patterns* by Robert C. Martin, Object Mentor, 2000

- *JEP-425: Virtual Threads*, `https://openjdk.org/jeps/425`

- *JSR-428: Structured Concurrency (Incubator)* (`https://openjdk.org/jeps/428`)

- *Patterns of Enterprise Application Architecture* by Martin Fowler, Pearson Education, Inc, 2003

- *Effective Java, Third Edition* by Joshua Bloch, Addison-Wesley, 2018

- *JDK 17: Class Exchanger* (`https://docs.oracle.com/en/java/javase/17/docs/api/java.base/java/util/concurrent/Exchanger.html`)

- *JDK 17: Class CompletableFuture* (`https://docs.oracle.com/en/java/javase/17/docs/api/java.base/java/util/concurrent/CompletableFuture.html`)

- *JDK 17: Class Executors* (`https://docs.oracle.com/en/java/javase/17/docs/api/java.base/java/util/concurrent/Executors.html`)

- *Java Mission Control* (`https://wiki.openjdk.org/display/jmc`)

Answers

1. The challenge solved by the double-checked singleton pattern is ensuring that only one class instance is present in the running JVM to avoid possible leaks

2. The usage of the `Executors` utility that resides in the `java.base` module and `java.util.concurrent` package

3. The balking pattern depends on the instance stat

4. The scheduler pattern

5. The producer-consumer pattern is one of the most common concurrent design patterns, with clearly separated and addressed logic

7

Understanding Common Anti-Patterns

Throughout the previous chapters, we have explored the *green working paths* of imaginary vehicle-related applications. In this chapter, the abstraction of the vehicle will remain a supporting element, because one may imagine a vehicle-inspired application more easily than other abstractions. A vehicle, along with all of its parts, is an idea that's easy to grasp.

Let's quickly recap the importance of design patterns and how they contribute to the success of an organization's success.

Melvin E. Conway said that the design and implementation of an application strongly reflect an organization's internal communication. This statement is no less relevant today, especially now that many projects use agile approaches. Automated builds, continuous integration or testing, and subsequent automated deployment play a key role in delivering applications to production. Any overlooked or unexpected limitation can limit or damage an application's main objective.

In this chapter, we will review some important areas for identifying signs of deviation from the main goal so that you can have functional, maintainable, and transparent applications. Not having those qualities can negatively affect application functionality on multiple levels. Runtime could be damaged, which might result in unpredictable costs. Faults could be hidden in application architecture, disallowing extendibility and maintainability. Such issues could also recur and require specialized attention in each case.

We will concentrate on the following areas:

- What anti-patterns are and how to identify them
- Examining typical software anti-patterns
- Understanding software architecture anti-patterns

By the end of this chapter, you will be able to identify and understand some of the signs of anti-patterns that should be looked out for.

Technical requirements

You can find the code files for this chapter on GitHub at `https://github.com/PacktPublishing/Practical-Design-Patterns-for-Java-Developers/tree/main/Chapter07`.

What anti-patterns are and how to identify them

One might define an anti-pattern as the exact opposite of a good design pattern or good practice. Though that might seem like the bottom line, it ignores the context and sequence of actions that led to a software practice being called an anti-pattern. In other words, it does not explain why an anti-pattern is a collection of highly risky, ineffective, and counterproductive steps. It is important to understand these steps as they allow creating a repeatable process to obtain a similar result, just to verify the ambiguity. The bottom line is that these steps may limit the ability to productively address the issues. Let us dive a bit deeper into the theory.

Theoretical principles challenges

Anti-patterns can naturally appear during software development due to multiple reasons. They may be due to a shift in business logic, technology migration, or missing information. The fact remains that anti-patterns do occur and, simply put, can be part of the development process due to team size, communication issues, and more.

The key question is how to identify them. In *Chapter 1*, *Getting into Software Design Patterns*, we touched on the negative impact of violating the APIE and SOLID design principles; this can be a symptom that can be a signal to consider a source code refactoring response. Another phenomenon would be not respecting the **CAP** theorem (**Consistency**, **Availability**, and **Partition tolerance**). Development time may have been invested in trying to achieve all three properties at the same time, which is impossible. Such attempts could be seen as a strong signal to rethink the development strategies.

Although the principles of APIE and SOLID tend to be considered common knowledge, the truth is different, especially in the area of creating agile approaches and completing tasks. Development can tend to create continuous technical debt. The word *continuous* is quite important because the accumulation of such debt may lead to very unpleasant consequences.

Collecting technical debt as a bottleneck

Technical debt is an interesting concept because it requires the context of the software application and its goal in order to understand it. Technical debt contains some properties that may not be obvious at first glance, but may lead to a serious application bottleneck. Let's imagine a vehicle production line as a collection of multiple processes running under various dynamics in parallel. The production line should deliver the expected result in the form of a vehicle. If there were an accumulation of bottlenecks, though, the result would not be achieved. This abstraction of a vehicle production line is quite simple, but putting it into the context of running software may be more difficult, as software relies on underlying technologies, platforms, and hardware.

The Java platform comes with some caveats that must be accounted for. Basically, the rule is that the developer respects the platform and the platform does its best to serve a running software.

Inappropriately squeezing the capabilities of the Java platform

In *Chapter 2, Discovering the Java Platform for Design Patterns*, we touched on important topics such as types and memory models; not accounting for these may be considered an anti-pattern. As a reminder of the multi-threaded nature of the Java platform, we can mention the Java memory model, which guarantees the visibility of values to the application in a multi-threaded environment. Another thing that may turn into an anti-pattern is a garbage collection algorithm running in a separate thread alongside the main application execution thread. Because a Java application works with the final allocated memory space (the heap), the garbage collector tries to ensure that there is still an appropriate amount of heap memory available for use.

The acquired knowledge helps us avoid one of the most common misconceptions relating to any anti-pattern: **unwanted autoboxing**, or the automatic conversion performed by the compiler between primitive types and the wrapper classes.

The impact of autoboxing is not visible at first glance and may go unnoticed until your application faces a critical load. Let us look at aggregating sensor values, where each sensor value needs to be validated to identify an alarm. The alarm is caused, of course, by the occurrence of a critical value. The alarm system starts multiple threads in parallel to verify the delivered values (*Example 7.1*):

```
record Sensor(int value) {}
class SensorAlarmWorker implements Runnable {

    . . .

    @Override
    public void run() {

        . . . .

        while (active) {

            . .

            Collection<Sensor> set = provider.values();
            for (Sensor e : set) {
                SensorAlarmSystemUtil.evaluateAlarm
                    (provider, e.value(), measurementCount);}

            . . .

    }
```

```
        }
    }
```

Example 7.1 – The SensorAlarmWorker instance attempts to identify a warning signal by reading the Sensor instance value

A vehicle sensor alarm system obviously has to analyze a huge amount of data delivered by its various sensors in order to identify a critical signal. An autoboxing issue tends to be very noticeable as it causes intensive and non-deterministic garbage collection (*Figure 7.1*):

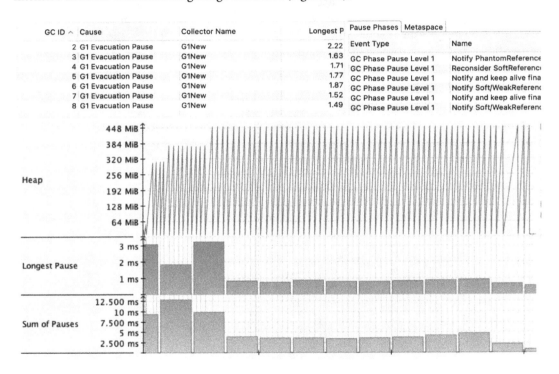

Figure 7.1 – The intensive garbage collection causes noticeable delays

Although the root cause of the garbage collection seems to be resolved, it may appear unexpectedly. *Example 7.1* introduces the `Sensor` record class, holding an integer value as a primitive `int` type. The problem becomes apparent when the value of the primitive type is autoboxed while passing it to the `evaluateAlarm` method, which requires the use of the `Integer` wrapper class. Let us do a one-line correction to the `Sensor` value type (*Example 7.2*):

```
....
static void evaluateAlarm(Map<Integer, Sensor> storage,
```

```
        Integer criticalValue, long measurementNumber) {
...
record Sensor(Integer value) {}
```

Example 7.2 – The Sensor field value is changed to the Integer type and corresponds with the method input type

This change has quite a significant impact on the whole application, causing very limited occurrences of garbage collection. In other words, eliminating unwanted delays due to stop-the-world events, as we learned in *Chapter 2, Discovering the Java Platform for Design Patterns* (*Figure 7.2*), speeds up the entire application:

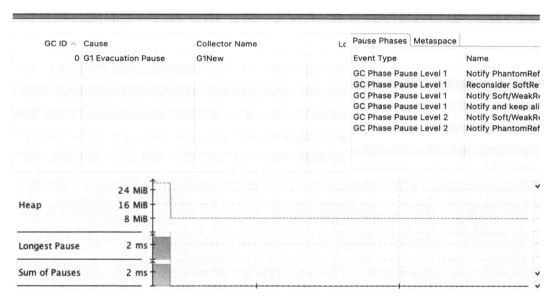

Figure 7.2 – Garbage collection pressure disappeared because the
application didn't create unnecessary short-living objects

Autoboxing may sometimes be identified through code reviews, which play an essential role in removing the code smell anti-pattern.

The Java platform contains many useful tools, and using them incorrectly can result in an unwanted state. Let's look at some of these tools in the next section.

Selecting the right tool

The next example may seem, at first glance, far from being a code smell. The Java platform contains very useful tools that can serve an application well if they are properly selected and used. A good example of the importance of careful selection is presented by the collection framework. *Chapter 2, Discovering the Java Platform for Design Patterns*, reviewed different aspects of commonly used collection types. This led us to see how a wrong collection selection may result in a bottleneck due to the consumption of underlying resources. This kind of issue may not be obvious with a small data amount but emerges under bigger loads and hits very specific parts of an application. This phenomenon can be called the **busy method**, or **hot method** (*Figure 7.3*):

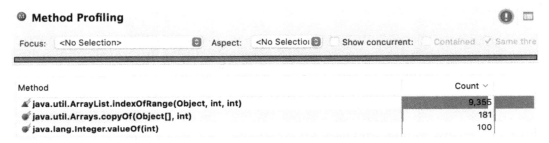

Figure 7.3 – The computation time is limited by busy method execution

Here the computation work of the application is centralized with one extremely busy method execution. **Java Flight Recorder** (**JFR**) has highlighted the issue. The fix became trivial because it rethought the use of the collection type. More specifically, it took the approach of accessing stored elements and replacing the O(n) time complexity for `ArrayList` with the O(1) time complexity for `HashSet` instances (*Figure 7.4*):

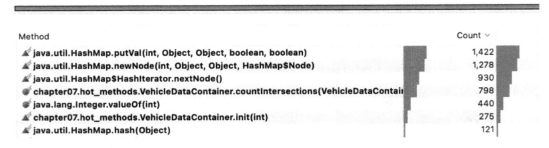

Figure 7.4 – The collection exchange leads to the desired computation work distribution

The correction results in bigger application throughput, which would be desirable for a vehicle data analyzer that attempts to evaluate collected data on a journey.

Although the code smell was not obvious, it was found because we used the right tool. Let's summarize what we can take from the previous sections.

Conclusion of the code smell anti-pattern

In this chapter, we discovered that any attempt to challenge theoretical principles leads to an anti-pattern known as a code smell.

We reviewed a case where unknown code smells posed a threat to an application's goals. We talked about how each bottleneck needs to be understood in its context before we try to solve it. Removing such bottlenecks without understanding the important details may result in another anti-pattern and a never-ending cycle of refactoring, as the presented examples showed (*Figure 7.1* and *Figure 7.3*).

Another counterproductive thing to do would be optimizing the code base without solving the underlying technical debt, which may result not only in questionable application throughput degradation but also unwanted behavior.

Before moving on to some of the most well-known software anti-patterns, let's make some final considerations. Any anti-pattern can be caused by a migration where technical debts have been created, including incorrect information. Their appearance may be a side effect of incorrectly selected platform tools or a lack of awareness of the theory behind the Java programming language. Since this topic can become controversial, I'll leave you to draw your own conclusions.

Examining typical software anti-patterns

The literature on this topic is full of different kinds of anti-patterns, some of which have very funny names, even though their impact is anything but funny. Sometimes anti-patterns can be the result of a lack of discipline in providing tested, well-structured, and maintainable code to colleagues. An often-used term today in this field is **clean code**. The following sections will explore some common anti-patterns that can be found in code bases, more specifically in method implementations.

Spaghetti code

Multiple factors may contribute to an application code base appearing very unstructured: that's the first sign of a code smell. In such cases, one of the most famous anti-patterns, **spaghetti code**, tends to appear. Spaghetti code may remain overlooked due to the fact that interfaces still remain coherent, but their implementation will contain long methods with interconnected dependencies (*Example 7.3*):

```
class VehicleSpaghetti {
    void drive(){
```

```
    /*
        around 100 lines of code
        heavily using the if-else construct
    */
    ...
```

Example 7.3 – An unclear drive() method that contains logic for everything from engine control to brake checks

In such a state, it is nearly impossible to extend the application or verify its functionality. Sometimes, such code may become legacy code, which people will then use as an excuse. Such excuses will not contribute to the application's success; the solution lies in refactoring and cleaning up the code base.

Cut and paste programming

This anti-pattern is perhaps another of the most common ones, where previously developed code is used to address the next challenge. This may seem like a smart reuse of code, but it can very quickly turn into a maintenance nightmare because the initial implementation conditions are completely ignored. This is a problem especially when the initial code was already anti-pattern-prone and ignored the principles mentioned in the previous chapters. The reuse of already-generated code should be done wisely in order to provide good protection.

Blob

This pattern can be identified in many older monolithic systems and applications. Nowadays, developers will claim that the problem has been overcome. The truth is not so certain. Not even frameworks are resistant to this anti-pattern. An anti-pattern can also be presented by a package that contains the most essential collection of divine classes or just one, the God class. This anti-pattern can often reside in classes called controllers, which control the entire behavior of an application. Such controllers accumulate a large number of different methods with different functionalities, meaning that the separation of concerns can end up being forgotten (*Example 7.4*):

```
class VehicleBlob {
    void drive(){}
    void initEngine(){}
    ...
    void alarmOilLevel(){}
    void runCylinder() {}
    void checkCylinderHead(){}
```

```
    void checkWaterPump(){}
}
```

Example 7.4 – The instance of VehicleBlob attempts to control each possible part

A counterargument regarding the hypothetical maintainability of the code base could be considered invalid because such code might be difficult to test, perhaps even impossible. A similar problem can occur when the singleton design pattern is improperly overused. Once the blob anti-pattern is identified, it's a great time to start thinking about creating simple diagrams supported by refactoring the source code before it's too late.

Lava flow

Clean code is sometimes used as a deep term. Nowadays, it is becoming a common practice to push a proof-of-concept application directly into production without further thought. This anti-pattern arises when incompatibility or extensibility problems start to appear. The fact that the proof of concept verified a possible solution does not guarantee that it was prepared for production according to common development principles and techniques. This anti-pattern can be identified by the occurrences of long implementation classes whose purpose has been lost over time, but everyone is afraid to remove them because it might affect the system. This anti-pattern was named after lava, a hot liquid that flows down a volcano until it catches fire. In the era of microservices, distributed systems, and cloud solutions, an example of shared functionalities, such as libraries or solutions, can be considered. When such a pattern emerges in the development process, it may be a good time to re-evaluate the design of the code base, perhaps draw some diagrams, and apply conclusions to mitigate the possibility of a fire.

Functional decomposition

The functional decomposition anti-pattern may seem outdated due to the use of modern frameworks and the fact that the anti-pattern is better known in the field of procedural languages. The reality may be slightly different, as many legacy systems were migrated without sufficient understanding of the code base and business logic. Identifying the anti-pattern is trivial, as it's impossible not to notice a code base containing many classes with a single responsibility, a lack of abstraction, and big cohesion. The root cause of this anti-pattern may be a lack of understanding of the basic principles of object-oriented programming or a misunderstanding of the application's goal. The solution is to refactor the code base according to the required abstraction level while keeping in mind programming principles.

Boat anchor

Sometimes an application or a newly developed piece of software may inherit an outdated abstraction that becomes unnecessary. This abstraction can turn into a bottleneck not only because it requires maintenance but also because it can easily be replicated widely across a code base. The worst case can be the large utilization of this abstraction inside shared libraries or application modules. The anti-pattern can accelerate the degradation of the application code base at various levels.

One easy way to mitigate this is to keep in mind the SOLID design and APIE principles to allow for continuous refactoring. This enables the utilization of previously learned design patterns.

Conclusion

Being able to identify and describe some of the most common deviations from well-known principles and approaches brings value to any project. In this subsection, we have examined anti-patterns and proposed solutions to keep your code base maintainable and readable. The last point we want to make in this section is about the proper naming of methods, fields, and classes, which can significantly improve readability and maintainability and limit the misunderstanding of API use. Proper naming is also important as it allows a good understanding of UML diagrams. The next section takes us more into source code architecture.

Understanding software architecture anti-patterns

A clear understanding of classes, packages, and module composition can be seen as essential not only to the application itself, but, as we learned in *Chapter 2, Discovering the Java Platform for Design Patterns*, to the platform as well. Thanks to the dynamic translation of the bytecode that the JIT compiler processes, the Java platform collects essential information about its optimization. Poor code quality and software architecture can cause latency, improper memory usage, or crashes. Let's understand the possible obstacles.

Golden hammer

One proven approach applied over a period of time without exploration of alternatives can easily turn into legacy code. The fact that it is difficult to accept other proposals or migration may be due to a particular set of approaches, otherwise known as the golden hammer anti-pattern, where developers believe that there is no need to investigate whether changes to something would be beneficial when it has been working fine for several years. A great example would be a vendor-specific database or tool, and the problem arises when the application needs to migrate to microservices or a more distributed design.

Scalability is penalized not only by the CAP theorem, as mentioned earlier, but may also suffer due to almost impossible usage across designed modules or application parts.

Using a particular vendor's product in your application architecture is not necessarily an issue. The challenge is that the development of the application relies entirely on the capabilities and functions provided by the vendor without evaluating its own capabilities.

A possible solution could be to re-evaluate current development approaches and allow for improvement through effective research of solutions.

Continuous obsolescence

Improvements are inevitable. Today, products can take advantage of automated deployment or continuous integration support with a variety of different test scenarios. The rate of improvement is increasing rapidly. A good example is the Java platform, which recently reduced its release cadence to 6 months. This fact can contribute to the emergence of an anti-pattern, since refactoring is required, but on the other hand, it has a great effect on removing previous shortcomings.

The continuous obsolescence anti-pattern can be easily identified by an inability to move a project to the next phase using continuous integration and delivery (*Example 7.5*):

```
interface VehicleCO {
    void checkEngine();
    void initSystem();
    void initRadio(); /* never used */
    void initCassettePlayer(); /* never used */
    void initMediaSystem(); /* actual logic */
}
```

Example 7.5 – The VehicleCO abstraction contains outdated methods that still need to be tested

Of course, the concepts of continuous delivery and integration do not give any guarantee of code base clarity as they need to be followed by development discipline. Continuous clean-code-focused reviews, object-oriented principles, and proper patterns can drastically reduce the incidence of continuous obsolescence and have a major impact on the entire application architecture.

Input kludge

It may not necessarily be obvious at first glance, but the input kludge anti-pattern is quite common. A good example would be several connected services that were tested until one began to deviate from its functionality. A quick ad hoc solution has more side effects that are recognized with a long delay due to disabled tests. Various services already had more patches applied than others, so even more tests were disabled. The fact remains that those disabled tests were critical to maintaining application integrity.

The solution might be to maintain discipline in testing and ensure that test inputs and outputs are valid and updated, rather than turning off essential tests.

Working in a minefield

Gone are the days of completely monolithic applications. The current distributed nature of applications creates an expectation for application testing to achieve continuous delivery and refactoring. Although an application might contain tests for known issues, full integrity cannot be ensured. What happens if some enhancements are made to an application but they aren't tested? Even a very small contribution to the code base can turn into a nightmare – working with such code bases can feel like being in a minefield. The solution to the problem is quite obvious: refactoring is required. Isolate the affected part with simplified tests to gain stability, apply all the knowledge learned, and slowly continue expanding the test base.

Ambiguous viewpoint

In the era of microservices and distributed system design approaches, this anti-pattern may significantly damage the final result. The continued presence of this anti-pattern can result in an onion architecture approach where the separation of concerns and other SOLID principles become theory rather than practice. One indicator of this anti-pattern would be the creation of unclear services followed by entities for passing redundant information between layers, making the general architecture unclear. It can be recognized in its early stages as the presented design models do not support any principles of SOLID application design due to incomplete information or unclear and overlapping perspectives in the model and potential software design. The solution may be effectively executed by the use of a modeling technique, such as UML, to ensure visual clarity and source code transparency.

Poltergeists

This anti-pattern cannot be fully ignored. It can be recognized if you can spot features that are not expected but suddenly appear and disappear. This anti-pattern is the result of very complex abstractions and the implementation of unnecessary classes. There are a couple of frameworks on the Java platform that can provide useless functionality. We can consider AspectJ and AOP as good examples as their usage may be directly responsible for mysterious side effects. The solution is to revisit and understand the class hierarchy and life cycles.

Dead end

The growth of the IT industry seems to be unstoppable, as more and more approaches and procedures for technical improvements emerge. As such, dependencies on non-updated components that were previously built into a system's architecture can be detrimental and may be more difficult to remove than you might expect. Take, for example, Java version migrations, where staying with the old version is penalized not only by losing support but also by increasing maintenance costs. Trying to scale an application can also result in many challenges to do with application testability.

Even though this dead-end anti-pattern can be turned into an accepted software design, it is recommended to consider alternatives, because doing so may come with significant costs.

Conclusion

Not all of the anti-patterns listed are completely bad (see the golden hammer or dead end, for example), but generally, any anti-pattern should be re-evaluated before it is accepted and documented.

Java is a very powerful language and platform, not only because it allows using instance mutation, but also because it enables entities to hold their state and remain immutable. Special attention should be paid to the state of the code base in the case of concurrent applications, because such applications must be transparent not only to software engineers but also to the Java platform, as we learned in *Chapter 2, Discovering the Java Platform for Design Patterns*. A concurrent environment brings many possibilities and performance improvements on the one hand, but on the other hand, it can cause the misuse of certain design patterns, such as the double-checked locking pattern, thanks to a poor understanding of the platform (rather than the framework that uses the platform).

A lack of test coverage, code writing discipline, information, or capability can also contribute significantly to the occurrence of anti-patterns. From a code base architecture perspective, verifying a correctly implemented function can become an impossible task or cause additional complications for mocking or debugging. Generally speaking, a transparent code architecture is one of the keys to a successful application. Let's recap everything we've learned.

Summary

Knowing about anti-patterns and their identification can have a big impact on the viability of an application, especially when it comes to distributed systems. This chapter has shown that not knowing the Java platform and its tools well enough can lead to unexpected results, such as intense pressure on garbage collection algorithms running on side threads and missing out on optimization opportunities. Realizing the multi-threaded nature of the Java platform can lead to proper code composition, but also to proper use of immutability to enable continuous application development, that is, refactoring.

Since most anti-patterns have a lack of testing as one of their root causes, a test environment might be the best starting place to determine the occurrence of anti-patterns or the need for refactoring. Compiled test code does not reside in deployed code, which makes test code the best place to start for future exploration and understanding your application's behavior.

The Java platform and other used libraries change rapidly, and one of the keys to maintaining an application code base is the correct use of the open-close principle. It enables continuous refactoring, which is essential for the evolution of a healthy code base.

Anti-patterns are part of the application life cycle. They are present and can remain present in various forms for the sake of progress. It may not be worth the effort to remove them entirely; perhaps a better approach would be to understand them and get the code base to the desired state with known limitations being addressed on an ongoing basis. Improvements will be made throughout the entire development cycle of the application, and writing code will become a pleasant experience with many challenges solved.

While the Java platform still has challenges, it remains a beautiful piece of software that uses mathematics, statistics, and probability science all at once!

Congratulations on successfully reaching the end of this book. Because every ending brings a new beginning, I encourage you to stay inspired and hungry and to have a lot of fun with coding or designing software! Keep your mind open and make it a source of valuable information.

Sincerely, Miro Wengner!

Further reading

- *Design Patterns: Elements of Reusable Object-Oriented Software* by Erich Gamma, Richard Helm, Ralph Johnson, and John Vlissides, Addison-Wesley, 1995

- *Design Principles and Design Patterns* by Robert C. Martin, Object Mentor, 2000

- *AntiPatterns: Refactoring Software, Architectures, and Project in Crisis* by William J. Brown, Raphael C. Malveau, Hays W. McCormick III, and Thomas J. Mowbray, John Wiley & Sons, Inc, 1998

- *CAP Twelve Years Later: How the "Rules" Have Changed*, `https://www.infoq.com/articles/cap-twelve-years-later-how-the-rules-have-changed`, 2012

- *Phoenix Project: A Novel About IT, DevOps, and Helping Your Business Win* by Gene Kim, Kevin Behr, and George Spafford, IT Revolution Press, 2016

- *How do Committees Invent?* by Melvin Edward Conway, Datamation 14, site 5, pages 28-31, 1968

- *Mission Control Project*, `https://github.com/openjdk/jmc`

Assessments

Chapter 1 – Getting into Software Design Patterns

1. The compiler compiles Java code into bytecode, which is executed by the JVM and JRE, respectively (refer to *Figure 1.3*).

2. It refers to abstraction, polymorphism, inheritance, and encapsulation.

3. Method overriding and method overloading.

4. SOLID principles: The single-responsibility principle, open-closed principle, Liskov substitution principle, interface segregation principle, and dependency inversion principle.

5. The program should be open for extension and closed for modification.

6. Design patterns represent the collection of commonly used problems and solutions to produce maintainable software.

Chapter 2 – Discovering the Java Platform for Design Patterns

1. The **Java Virtual Machine (JVM)**, **Java Runtime Environment (JRE)**, and **Java Development Kit (JDK)**.

2. Java is a statically typed language, which means any value needs to be declared before it can be assigned to the value.

3. 3. Primitive types: `boolean`, `byte`, `short`, `char`, `int`, `float`, `long`, and `double`.

4. Garbage collector.

5. `Queue`, `Set`, and `List`.

6. Key-value pairs.

7. In O-notation O(1).

8. In O-notation O(n).

9. `Predicate<T>`, the return type is a primitive type, `boolean`.

10. Element streams in the Java Stream API are lazily evaluated.

Chapter 3 – Working with Creational Design Patterns

1. Creational design patterns help abstract the object instantiation process by delegating it to the responsible part of the application.

2. 2. To reduce new object creation costs, the dependency injection, lazy initiation, and object pool patterns may be taken into consideration.

3. Only one instance is required to be present in JVM.

4. The builder pattern helps create configurations of a similar object type while reducing the number of constructors.

5. The factory method or abstract factory patterns should be considered, as both can compose complex objects without exposing the logic to the clients.

6. The object pool design pattern introduces a cache of already created and reusable objects instead of allocating and destroying new instances.

7. The most useful pattern for creating objects of a specific family is the factory method pattern.

Chapter 4 – Applying Structural Design Patterns

1. 1. Structural design patterns define communication between objects. These patterns support implementation flexibility and transparency.

2. The structural design patterns described by the GoF author group are the adapter, bridge, composite, proxy, flyweight, facade, and decorator patterns.

3. The composite structural design pattern, which also guarantees uniform object handling.

4. The marker pattern, with full awareness of its drawbacks.

5. The proxy pattern needs to be taken into account because the adapter and facade patterns have slightly different purposes.

6. The bridge pattern.

Chapter 5 – Behavioral Design Patterns

1. The Liskov substitution principle explored in *Chapter 1, Getting into Software Design Patterns*.

2. The iterator pattern.

3. Yes – the strategy pattern.

4. It is the null object pattern, which provides the type of such a state and limits the causes of null pointer exceptions.

5. This can be the pipeline pattern, the strategy pattern for the map() and filter() methods, or the null object pattern.

6. All clients can be alerted by employing the observer pattern, which also transparently controls conditions.

7. The command pattern can be used. A command is represented by a unique object. An object allows a client to pass parameters and can easily call a callback function.

Chapter 6 – Concurrency Design Patterns

1. The challenge solved by the double-checked singleton pattern is ensuring that only one class instance is present in the running JVM to avoid possible leaks

2. The usage of the `Executors` utility that resides in the `java.base` module and `java.util.concurrent` package

3. The balking pattern depends on the instance stat

4. The scheduler pattern

5. The producer-consumer pattern is one of the most common concurrent design patterns, with clearly separated and addressed logic

Index

www.packtpub.com

Subscribe to our online digital library for full access to over 7,000 books and videos, as well as industry leading tools to help you plan your personal development and advance your career. For more information, please visit our website.

Why subscribe?

- Spend less time learning and more time coding with practical eBooks and Videos from over 4,000 industry professionals

- Improve your learning with Skill Plans built especially for you

- Get a free eBook or video every month

- Fully searchable for easy access to vital information

- Copy and paste, print, and bookmark content

Did you know that Packt offers eBook versions of every book published, with PDF and ePub files available? You can upgrade to the eBook version at packtpub.com and as a print book customer, you are entitled to a discount on the eBook copy. Get in touch with us at customercare@packtpub.com for more details.

At www.packtpub.com, you can also read a collection of free technical articles, sign up for a range of free newsletters, and receive exclusive discounts and offers on Packt books and eBooks.

Other Books You May Enjoy

If you enjoyed this book, you may be interested in these other books by Packt:

Domain-Driven Design with Java - A Practitioner's Guide

Premanand Chandrasekaran, Karthik Krishnan

ISBN: 9781800560734

- Discover how to develop a shared understanding of the problem domain
- Establish a clear demarcation between core and peripheral systems
- Identify how to evolve and decompose complex systems into well-factored components
- Apply elaboration techniques like domain storytelling and event storming
- Implement EDA, CQRS, event sourcing, and much more
- Design an ecosystem of cohesive, loosely coupled, and distributed microservices
- Test-drive the implementation of an event-driven system in Java
- Grasp how non-functional requirements influence bounded context decompositions

Java Memory Management

Maaike van Putten, Seán Kennedy

ISBN: 9781801812856

- Understand the schematics of debugging and how to design the application to perform well
- Discover how garbage collectors work
- Distinguish between various garbage collector implementations
- Identify the metrics required for analyzing application performance
- Configure and monitor JVM memory management
- Identify and solve memory leaks

Packt is searching for authors like you

If you're interested in becoming an author for Packt, please visit `authors.packtpub.com` and apply today. We have worked with thousands of developers and tech professionals, just like you, to help them share their insight with the global tech community. You can make a general application, apply for a specific hot topic that we are recruiting an author for, or submit your own idea.

Share Your Thoughts

Now you've finished *Practical Design Patterns for Java Developers*, we'd love to hear your thoughts! Scan the QR code below to go straight to the Amazon review page for this book and share your feedback or leave a review on the site that you purchased it from.

`https://packt.link/r/1-804-61467-X`

Your review is important to us and the tech community and will help us make sure we're delivering excellent quality content.

Download a free PDF copy of this book

Thanks for purchasing this book!

Do you like to read on the go but are unable to carry your print books everywhere? Is your eBook purchase not compatible with the device of your choice?

Don't worry, now with every Packt book you get a DRM-free PDF version of that book at no cost.

Read anywhere, any place, on any device. Search, copy, and paste code from your favorite technical books directly into your application.

The perks don't stop there, you can get exclusive access to discounts, newsletters, and great free content in your inbox daily

Follow these simple steps to get the benefits:

1. Scan the QR code or visit the link below

https://packt.link/free-ebook/9781804614679

2. Submit your proof of purchase

3. That's it! We'll send your free PDF and other benefits to your email directly

www.ingramcontent.com/pod-product-compliance
Lightning Source LLC
Chambersburg PA
CBHW060534060326
40690CB00017B/3486